The NASA STI Program Office ... in Profile

Since its founding, NASA has been dedicated to the advancement of aeronautics and space science. The NASA Scientific and Technical Information (STI) Program Office plays a key part in helping NASA maintain this important role.

The NASA STI Program Office is operated by Langley Research Center, the lead center for NASA's scientific and technical information. The NASA STI Program Office provides access to the NASA STI Database, the largest collection of aeronautical and space science STI in the world. The Program Office is also NASA's institutional mechanism for disseminating the results of its research and development activities. These results are published by NASA in the NASA STI Report Series, which includes the following report types:

- TECHNICAL PUBLICATION. Reports of completed research or a major significant phase of research that present the results of NASA programs and include extensive data or theoretical analysis. Includes compilations of significant scientific and technical data and information deemed to be of continuing reference value. NASA counterpart of peer-reviewed formal professional papers, but having less stringent limitations on manuscript length and extent of graphic presentations.

- TECHNICAL MEMORANDUM. Scientific and technical findings that are preliminary or of specialized interest, e.g., quick release reports, working papers, and bibliographies that contain minimal annotation. Does not contain extensive analysis.

- CONTRACTOR REPORT. Scientific and technical findings by NASA-sponsored contractors and grantees.

- CONFERENCE PUBLICATION. Collected papers from scientific and technical conferences, symposia, seminars, or other meetings sponsored or co-sponsored by NASA.

- SPECIAL PUBLICATION. Scientific, technical, or historical information from NASA programs, projects, and missions, often concerned with subjects having substantial public interest.

- TECHNICAL TRANSLATION. English-language translations of foreign scientific and technical material pertinent to NASA's mission.

Specialized services that complement the STI Program Office's diverse offerings include creating custom thesauri, building customized databases, organizing and publishing research results ... even providing videos.

For more information about the NASA STI Program Office, see the following:

- Access the NASA STI Program Home Page at *http://www.sti.nasa.gov*

- E-mail your question via the Internet to help@sti.nasa.gov

- Fax your question to the NASA STI Help Desk at (301) 621-0134

- Phone the NASA STI Help Desk at (301) 621-0390

- Write to:
  NASA STI Help Desk
  NASA Center for AeroSpace Information
  7115 Standard Drive
  Hanover, MD 21076-1320

NASA/TM-2008-215126/Vol I
NESC-RP-06-108/05-173-E

Design Development Test and Evaluation (DDT&E) Considerations for Safe and Reliable Human Rated Spacecraft Systems

James Miller
NASA Langley Research Center, Hampton, Virginia

Jay Leggett
NASA Langley Research Center, Hampton, Virginia

Julie Kramer-White
NASA Johnson Space Center, Houston, Texas

NASA Engineering and Safety Center
Langley Research Center
Hampton, Virginia 23681-2199

April 2008

The use of trademarks or names of manufacturers in the report is for accurate reporting and does not constitute an official endorsement, either expressed or implied, of such products or manufacturers by the National Aeronautics and Space Administration.

Available from:
NASA Center for AeroSpace Information (CASI)
7115 Standard Drive
Hanover, MD 21076-1320
(301) 621-0390

	NASA Engineering and Safety Center Technical Report	Document #: RP-06-108	Version: 1.0
	Design, Development, Test, and Evaluation (DDT&E) Considerations for Safe and Reliable Human Rated Spacecraft Systems		Page #:

Volume I

Design Development Test and Evaluation (DDT&E) Considerations for Safe and Reliable Human Rated Spacecraft Systems

May 1, 2007

NESC Request Number: 05-173-E

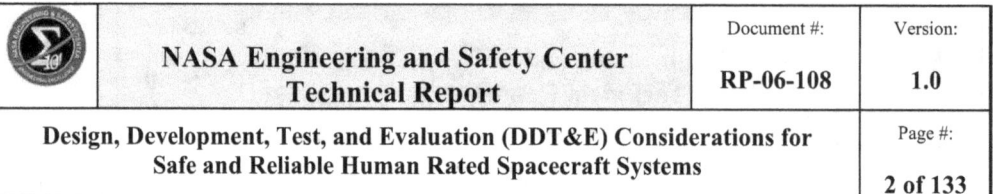

		Document #:	Version:
	NASA Engineering and Safety Center Technical Report	RP-06-108	1.0
	Design, Development, Test, and Evaluation (DDT&E) Considerations for Safe and Reliable Human Rated Spacecraft Systems	Page #: 2 of 133	

Approved:	Original signed on file	5-3-07
	NESC Director	Date

Revision	Description of Revision	Author	Effective Date
Base	Initial Release	NESC Principal Engineers Office	12/07/06

NESC Request Number: 05-173-E

NASA Engineering and Safety Center Technical Report	Document #: RP-06-108	Version: 1.0
Design, Development, Test, and Evaluation (DDT&E) Considerations for Safe and Reliable Human Rated Spacecraft Systems		Page #: 3 of 133

Table of Contents

Volume I: Consultation Report

Signature Page 6
Executive Summary 7
1.0   Introduction 27
        1.0.1   Definitions 30
    1.1   Historical Context of Space Systems Safety and Reliability 31
    1.2   Examination of Failure History Data 34
2.0   System Engineering with a Safety and Reliability Focus 44
    2.1   The Right Work at the Right Time with the Right Teams 47
        2.1.1   Defining Program Needs, Objectives, and Constraints 47
        2.1.2   Organizing and Managing the Program 48
        2.1.3   Planning and Pacing of Work throughout the Life Cycle 49
        2.1.4   Teamwork for Producing Safe and Reliable Systems 51
    2.2   Defining the Requirement for a Safe Human Rated System 54
        2.2.1   Fault Tolerant Requirement 57
        2.2.2   Quantitative Requirement and Supporting Analysis 60
    2.3   Conceiving the Right System, Critical Activities Early in the Life Cycle 62
        2.3.1 Managing Complexity 63
        2.3.2   Iterative System Design and Defining the Right Requirements 65
        2.3.3   Risk Based System Design Loop 68
        2.3.4   System Redundancy Design Guidance or Rules of Thumb 77
        2.3.5   Lessons from the real world: 78
    2.4   Implementing the System Right, Achieving a Safe and Reliable System 80
        2.4.1   Design 82
        2.4.2   Manufacturing, Assembly, and Integration 86
        2.4.3   Independent Review 87
        2.4.4   Inspection and Walkdown 90
        2.4.5   Product Verification and Validation, "Test Like You Fly" 90
        2.4.6   Operate, "Fly Like You Test" 93
    2.5   Integrating Risk 94
        2.5.1   Identifying and Classifying Risks 94
        2.5.2   Incremental Acceptance of Risk 98
        2.5.3   Evaluating and Trading Disparate Risks 98
        2.5.2   Integrating Cost Risk Performance Model into CRM 100
3.0   Safety and Reliability Analysis throughout the Life Cycle 102
    3.1   Formulation Phases 104
        3.1.1   Pre-Phase A Advanced Studies 104
        3.1.2   Phase A Preliminary Analysis 106

NASA Engineering and Safety Center Technical Report	Document #: RP-06-108	Version: 1.0
Design, Development, Test, and Evaluation (DDT&E) Considerations for Safe and Reliable Human Rated Spacecraft Systems		Page #: 4 of 133

    3.1.3   Phase B Definition 109
  3.2   Implementation Phases 111
    3.2.1 Phase C Design 111
    3.2.2 Phase D Development... 112
    3.2.3 Phase E Operation 113
  3.3   Application of Risk Analysis Tools and Techniques...................................... 114
  3.4   Key Issues in Quantifying Risk 124
    3.4.1 Reliability Analysis Consideration of Common Cause Failures 124
    3.4.2 Maturity Modeling 126
    3.4.3 Heritage Data 127
Acronyms.. 128

List of Figures

Figure i-1	Three Pronged Safety and Reliability Requirement 12
Figure i-2	Complexity versus Development Time and Cost 14
Figure i-3	Project Constraints Box Showing Alternatives as a Surface with Selected Solution 17
Figure i-4	Objectives Driven and Risk Based Iterative Design Loop 20
Figure i-5	Multilayered Approach to Produce a Safe and Reliable System and Screen for Potential Problems... 22
Figure i-6	Risk Information Flow.. 24
Figure i-7	Guiding Principles Applied to End-to-End Development of a Safe and Reliable System 26
Figure 1.2-1	Significant Human Space Vehicle Failures 35
Figure 1.2-2	Individual Parts Failure Rates 36
Figure 1.2-3	Test Assessment and Risk Management... 37
Figure 1.2-4	A Complexity-Based Risk Assessment of Low-Cost Planetary Missions: When is a Mission Too Fast and Too Cheap 38
Figure 1.2-5	A Successful Strategy for Satellite Development and Testing 39
Figure 1.2-6	Flight Failures with Software as Major Cause... 40
Figure 1.2-7	Orbital Experience from an Integration and Test Perspective 41
Figure 1.2-8	Reliability Prediction for Spacecraft... 41
Figure 1.2-9	Analysis from Aerospace Corporation Space Systems Engineering Database (SSED) 42
Figure 1.2-10	Demonstrated Launch Vehicle Reliability Improvement with Maturity,............. 43
Figure 2.0-1	Conceiving the Right System and Implementing the System Right.................... 44
Figure 2.0-2	Trade Space and Constraint "Box".. 46
Figure 2.1-1	Planning and Pacing of Work 50
Figure 2.1-2	Cost Committed versus Cost Incurred 51
Figure 2.1-3	Multilayered Approach to Produce a Safe and Reliable System 53

Figure 2.2-1	Three Pronged Safety and Reliability Requirement	57
Figure 2.3-1	Early Systems Design Flow	67
Figure 2.3-2	Iterative Risk Based System Design Loop	69
Figure 2.3-3	Event Sequence Diagram Example	75
Figure 2.3-4	ESAS Variation of Mission Architectures with LOC Risk	76
Figure 2.3-5	ESAS Probability of LOC Variation with CLV LEO Configuration	77
Figure 2.4-1	Multilayered Approach to Produce a Safe/Reliable System and Screen for Hazards	81
Figure 2.4.2	Candidate Design and Assessment Iterative Detailed Design Loop	83
Figure 2.5-1	Consequence Focused Risk Types	95
Figure 2.5-2	Risk Information Flow	97
Figure 2.5-3	Qualitative, Quantitative, and Probabilistic Assessment Methods	99
Figure 2.5-4	Life Cycle Risk Profile	100
Figure 3.1-1	Systems Engineering Life Cycle with Reliability Focus	103
Figure 3.4-1	Conditional Probabilities of Multiple Failures of Identical Components	125
Figure 3.4-2	Effects of Redundancy and Dependent Failures on Reliability	126
Figure 3.4-3	Demonstrated Launch Vehicle Reliability Improvement with Maturity	127

List of Tables

Table 1.1-1	Comparison of Historical Reliability Approaches	33
Table 2.2-1	Probability of Loss of Aircraft Summary Data	61
Table 3.3-1	Reliability Assessment Functions Along the Systems Life Cycle	117
Table 3.3-2	Reliability and Risk Assessment Tools and Techniques	120
Table 3.3-3	Information Required for Reliability Assessments	122

Signature Page

Original signatures on file

_____   _____
Mr. Jay Leggett   Date   Mrs. Julie Kramer White   Date

_____   _____
Mr. James Miller   Date   Dr. Bernard Adelstein   Date

_____   _____
Mr. Michael Aguilar   Date   Mr. Michael Bay   Date

_____   _____
Mr. Mitchell Davis   Date   Mr. Cornelius Dennehy   Date

_____   _____
Mr. George Hopson   Date   Mr. John McManamen   Date

_____   _____
Dr. Bob Piascik   Date   Dr. Ivatury Raju   Date

_____
Mr. Hank Rotter   Date

NESC Request Number: 05-173-E

Executive Summary

With the launch of the Constellation program, NASA finds itself with the opportunity to design the next generation of manned vehicles that will take man to the Moon and beyond in the next two decades. While there are precedents for many aspects of the design, development, test, and evaluation (DDT&E) task at hand – the Apollo program, Space Transportation System, International Space Station (ISS) and others – the Johnson Space Center (JSC) Astronaut Office asked for a fresh look at first principles during the formative times of this new program.

As a result, a team directed by the NASA Engineering and Safety Center (NESC) collected methodologies for how best to develop safe and reliable human rated systems and how to identify the drivers that provide the basis for assessing safety[1] and reliability. The team also identified techniques, methodologies, and best practices to assure that NASA can develop safe and reliable human rated systems. The results are drawn from a wide variety of resources, from experts involved with the space program since its inception to the best-practices espoused in contemporary engineering doctrine. This report focuses on safety and reliability considerations and does not duplicate or update any existing references. Neither does it intend to replace existing standards and policy.

Summary of Fundamental Top Level Conclusions

A. History indicates no subsystem, component, or system element is immune from failure. While systems and their component parts have become more mature and reliable over the last four decades, the increase in system capability brings with it complexity and uncertainty in performance that can threaten any system element. The most frequent failure causes are in design, understanding of the environment, parts, and workmanship. The potentially generic, cross cutting, nature of these causes threatens the system, its redundancy, and other hazard controls.

B. There is no single requirement, method, or process, which, by itself, assures the "right stuff" for safety and reliability. The realities of the complex and uncertain nature of space flight manifested throughout the life cycle require a multifaceted approach to controlling risk.

C. System level safety and reliability is achieved by maintaining a focus on these throughout the DDT&E life cycle and during spacecraft operations. Safety and mission success assessments should be performed by the design team periodically throughout the life cycle as the design matures into hardware, software and a flight system. Assessments start from a top down functional perspective at the beginning of the life cycle and then mature along with the system design to include component design details. By focusing the design team on safety, and adherence to basic principles, the development team will seek to predict how the system will perform, discover safety and mission success risks, and obviate or mitigate their consequences before the system becomes operational. The designers of the system are

---

[1] "Safety" as used in this report (Section 1.0.1) restricts the definition to personnel loss or injury, in contrast to the broader definition used in Mil-Std 882 which includes loss of mission and loss of hardware.

ultimately responsible for the safety of the design, and the analysis to demonstrate that safety; safety, reliability, and quality assurance organizations and personnel are there to facilitate this process and provide independent oversight.

Guiding principles:

1. Define a clear and simple set of prioritized program needs, objectives and constraints, including safety, that form the validation basis for subsequent work.
2. Manage and lead the program with a safety focus, simple and easy to understand management structures, and clear lines of authority and responsibility among the elements.
3. Specify safety and reliability requirements through a triad of fault tolerance, bounding failure probability, and adhering to proven practices and standards.
4. Manage complexity by keeping the primary (mission) objectives as simple and minimal as possible and adding complexity to the system only where necessary to achieve these objectives.
5. Conceive the right system conceptual design early in the life cycle by thoroughly exploring risks from the top down and using a risk-based design loop to iterate the operations concept, the design, and the requirements until the system meets mission objectives at minimum complexity and is achievable within constraints.
6. Build the system right by applying a multilayered, "defense in depth"[2] approach of following proven design and manufacturing practices, holding independent reviews, inspecting the end product, and employing a "test like you fly, fly like you test" philosophy.
7. Seek and collect warning signs and precursors to safety, mission success and development risks throughout the life cycle, and integrate those into a total risk picture with appropriate mitigation activities.

These principles are supported by a foundation of established project management, systems engineering, safety and mission assurance, and operations practices that encourages a safety focus throughout the program life cycle. The safety focus includes the attitude and approach to safety of those conceiving, producing and operating the integrated system in addition to individual components and system elements.

---

[2] Dunn, M, Remaking NASA one step at a time, Associated Press, October 12, 2003, ("The opinions of technicians and engineers, no matter how low on the ladder, were not only respected, but sought by flight directors like the legendary Kranz. He practiced "defense in depth," so that if a technical problem slipped past one group, it would be caught by the next, or the next. He demanded toughness, competence, confidence.")

Safety Focus

When considering the safety of a design, it is critical to maintain a top-down view of the system as a whole and answer questions like: "What can go wrong to affect safety?", "What are the consequences?", and "How likely is it?" These questions have equal bearing as performance and cost, especially when considering critical systems and functions. Assuring the correct questions are asked requires experience, understanding of the mission objectives and a sense of curiosity, skepticism, and imagination for what could go wrong.

Next, the team must answer these questions and identify steps to make the total system safe and reliable so ultimately the project can affirmatively state, "The system is safe to fly because ___ was done." By the same token, weak links must be addressed as, "It will be safe if and when ___ is done."

The safety focus seeks to obviate or mitigate risks before they surface as problems in flight. This is achieved by developing an understanding of how the integrated system is expected to perform and how it might fail through a thorough examination of its design and its use.

Distinguishing between Safety and Mission Success

This report distinguishes between safety and mission success to provide a consequence-based focus that is useful during design, production, operations, and overall risk management. Distinguishing between safety and mission success is important when deciding when the mission should be aborted to preserve crew safety and when efforts to preserve mission success may adversely impact safety.

A safe system ensures the survival, health, and well-being of the crew during nominal and off-nominal operational scenarios. It also provides strategies to avoid or deal with unsafe conditions and applies margin to the system to prevent the exceedance of limits that may result in harm to the crew.

A reliable system assures mission success by functioning properly over its intended life. It has a low and acceptable probability of failure, achieved through simplicity, proper design, and proper application of reliable parts and materials. In addition to long life, a reliable system is robust and fault tolerant, meaning it can tolerate failures and variations in its operating parameters and environments.

Safety and reliability objectives often work together, but can also compete. The two work together when margins are added to the system to ensure its continued operation from both safety and mission success perspectives. They compete when safety objectives seek to prevent a hazardous condition that also interrupts mission success. For example, a human rated system would set safety limits, such as engine red lines, prior to the point of failure to allow a crew abort scenario. An early or false abort may occur at the expense of mission success.

The team producing this report highlighted seven guiding principles for DDT&E based upon nearly five decades of space travel that drive safety and reliability. A discussion of each item is presented below and further developed later in the report:

1. Defining Program Needs, Objectives, and Constraints (Section 2.1)

Define a clear and simple set of prioritized program needs, objectives, and constraints, including safety, that form the validation basis for subsequent work.

Primary requirements of the spacecraft and its related systems are derived from the mission objectives. The more simply and easily the mission objectives are to define, the easier the job of validating the requirements and system. A senior program manager in the Apollo days, interviewed after the successful conclusion of the program, said, "The mission was easy to define: Man, the Moon, this Decade." Thanks to the clarity of that overarching goal, he was always able to test the need for any proposed new requirements by asking, "Does this help get a man on the Moon in this decade and return him to safely home?" If the answer was no then the design would continue without it.

However, if the mission objective is expansive and ill-defined, there is room for all manner of add-ons that could eventually produce a system that either cannot be built or, if built, cannot be operated with a high degree of confidence. Consider the guidance during the initial days of the Space Transportation System (STS) development which called for a versatile system capable of being adapted to perform all plausible, but undefined, future missions. With such open-ended guidance, practically any new requirement can be defended, leading the design to grow in complexity.

Most programs have mission objectives that are somewhere between the two extremes, often with room for a variety of interpretations. As a result, the programs can become subject to mission creep that can lead to schedule slips and increased weight, cost, and complexity, the antithesis of reliability.

2. Organizing and Managing the Program (Section 2.1)

Manage and lead the program with a safety focus, simple and easy to understand management structures, and clear lines of authority and responsibility among the elements.

Having established an appropriate objective, the next most important step is to create an organizational structure that can carry out the program. The organization should be staffed with experienced people who understand the needs and objectives, can conceive and produce a system that meets those needs, and have the curiosity, skepticism, and imagination to identify what can go wrong, and can then take appropriate action. This must go beyond asking the question, "Am I meeting requirements?"

The most effective means of producing successful systems is to put properly prepared and professionally competent managers in charge, provide them with adequate resources, challenge them with a difficult but doable schedule, and give them sufficient authority to be held personally responsible for results. Experience shows that competent people who accept personal responsibility for the final outcome will do more to find the right balance among the plethora of competing demands that must be addressed during development.

Executive management has three critical responsibilities in making this process work.

- Selecting the right objectives, organizational structure, and key personnel is first.
- After selecting the right people, they must ensure that each manager's authority and responsibility are unambiguous, co-aligned, and supported.
- Finally, having delegated authority and responsibility to a team that will be working under intense pressure for years, management must provide a safety net by maintaining discipline in execution of the Program Plan, ensuring transparency, and providing timely independent assessments of progress. Such activities are not bureaucratic but enable the team to operate aggressively while knowing there will be a second set of eyes to help them avoid errors of commission and omission. Through formal reviews and informal communications the executive leadership will gain insight into the state of the program and health of the team that is executing it. In so doing, the probability of identifying and correcting the inevitable mistakes will increase, thereby protecting the team, mission, and agency.

The organization should support and encourage a safety focus among team members including suppliers. The effectiveness of the safety focus depends on the following factors depicted as the poles in Figure i-5:

- The team's experience and their access to experienced personnel when needed,
- Their understanding and acceptance of the mission objectives,
- A sense of curiosity, skepticism, and imagination to identify what might go wrong. The team should ask, "Why is it safe?" throughout the program life cycle. Affirmative answers to these questions create a preponderance of evidence the system is safe to fly. Concluding an affirmative in the absence of contrary evidence ("It has never failed before") must be avoided.[3] Long strings of successes can unintentionally lead to misunderstanding or underestimating risks inherent in the system, and
- Free and open discussion of safety issues among all team members and suppliers is critical to assuring that all parts of the system receive consistent attention to safety.

---

[3] Kraft, C., Flight, My Life in Mission Control, (Dutton, 2001), p 98 ("If somebody says that something never happens, be prepared because it probably will"), (This book provides good insight into the degree that the early manned spaceflight era focused on expecting the unexpected and not taking success for granted. They focused on making the unknowns known and being prepared should one surface)

3. Defining Safety and Reliability for Human Rated Systems (Sections 2.2 and 3.0)

Specify safety and reliability requirements through a triad of fault tolerance, bounding failure probability, and adhering to proven practices and standards.

A safe and reliable human rated spacecraft system provides a reasonable assurance that the crew survives nominal and off-nominal operational scenarios. NASA-STD-3000 Man-Systems Integration Standards provides a set of requirements that assure crew survival in the space environment including a breathable atmosphere, safe thermal and g-loads, safe radiation environment, and adequate nourishment and personal health and hygiene opportunities.

Assurance of crew survival during off-nominal and failure scenarios leads to requirements that seek to prevent faults and assure crew safety in spite of faults. To maximize safety and reliability, design the system with a fault tolerant architecture supported by probabilistic safety, reliability, and risk analyses backed up by data, hard evidence and/or analysis whenever possible. The system should then be manufactured and operated according to proven practices. Taken together, these three elements (fault tolerance, bounding probability of failure, and adhering to proven practices) as shown in Figure i-1, form a triad that defines the overall safety and reliability requirements.

Assessing safety and reliability starts at the onset of the program with high level functions and continues throughout the life cycle as the design matures and system elements are produced, integrated, and prepared for flight.

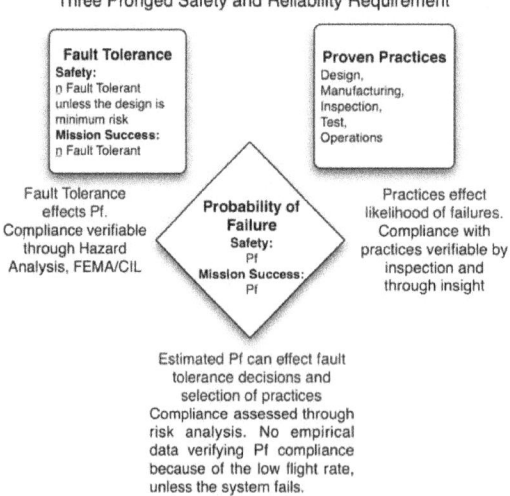

Figure i-1 Three Pronged Safety and Reliability Requirement

- Assess fault tolerance at the program level from a top down functional perspective and flow it to lower levels only after it is optimized from the total system perspective. For example, levying a two fault tolerance requirement at lower levels may introduce system complexities that may be inappropriate since mitigation can occur at higher levels where the systems interact.

- The objective of bounding the probability of failure is to encourage a thorough investigation into risks including uncertainty and common cause such that the system design decisions and underlying risk analysis can be defended. Estimating the likelihood of failure challenges the design team to achieve a deeper understanding of the system and its environment and provides a way to make trades when protecting against worst case assumptions that cause a design to become overly complex or over weight. The value of a probability estimate is not so much contained in the absolute number but in the thorough investigation, debate, and discussions by designers and operators about controlling the potential for failures based on their likelihood, history of similar systems, and uncertainties inherent in the system design. The analysis of the system design must consider the integrated whole and include a top down assessment. The analyses are most useful for evaluating and comparing design and operations alternatives and validating the chosen system design. To ensure valid comparison correct statistical methods should be used to determine probability of failure and include all available data sources.

- Identify and follow proven practices and processes for design, manufacturing, independent review, inspection, test, and operations. Processes can achieve consistency and provide a method for following through on details necessary for safety and reliability. Disciplined adherence to critical processes is important for achieving the safety and reliability the designers intended.

4. Managing Complexity (Section 2.0)

Manage complexity by keeping the primary (mission) objectives as simple and minimal as possible and adding complexity to the system only where necessary to achieve these objectives.

System complexity must be minimized since it is the most significant feature of systems that fail. Complexity impedes the designer's understanding of how various system elements might interact and could prevent a full understanding of the integrated system. Human spaceflight operates on the boundaries of technological capabilities. It is a highly integrated activity that is complex and requires the sequential success of a large number of active subsystems all of which are operating close to their limits. As such, a small increase in complexity may have a negative impact on safety and reliability. Consider the following graphs in Figure i-2 that show complexity as a function of cost and development time for a wide range of spacecraft, some successful, some not.

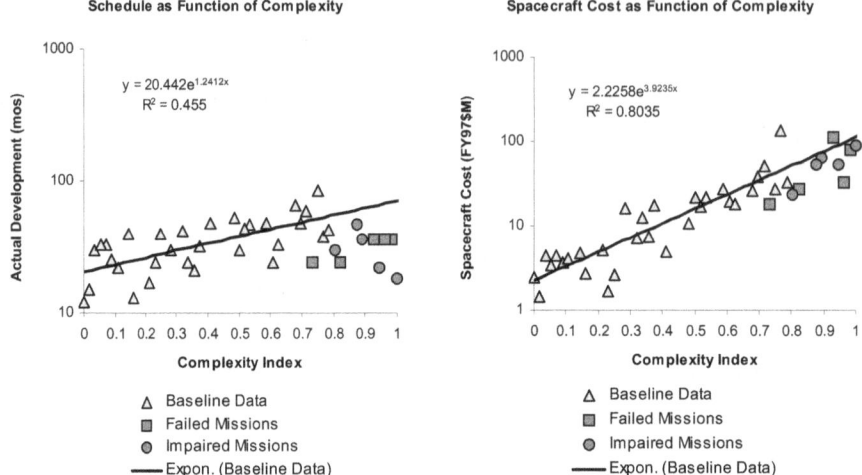

Figure i-2 Complexity versus Development Time and Cost

One way to interpret the graphs is that a complex mission must be provided an adequate budget and sufficient time to be successful. Another interpretation suggests that if a mission is sufficiently complex, it will probably fail. Regardless of the interpretation, complexity is the antithesis of reliability and should be limited to what is needed to accomplish the mission objective.

Managing and integrating pieces into a cohesive whole. A common method for managing large and complex systems is to divide the whole into smaller, simpler "manageable" pieces, and then allow separate groups to individually produce those pieces. The splitting of the system into pieces must occur from the top down considering the critical functions necessary for safety and reliability. Engineering managers must have a firm grasp of the risk drivers for their system even if they are at very low levels of the Work Breakdown System (WBS).

The engineering and management challenge then becomes the process of reintegrating the pieces into a cohesive system while avoiding adverse couplings and interactions that may affect safety and reliability.

Simplifying design, modeling, and interfacing assumptions early in the life cycle often turn out to be more complex when actual systems are produced and tested. The role of the systems engineer in integration requires the mindset of a "generalist" who can identify critical functional, physical interfaces, and interactions among tightly coupled system elements. Functional and physical interfaces must be kept simple so he can identify where newly joined elements of the system may adversely interact and compromise safety and reliability.

Design teams responsible for individual system elements must be aware of their system's sensitivities and unwanted interactions with other system elements to understand potential adverse coupling with other systems. It is important for the Systems Engineering Team to recognize the importance of interaction among discipline engineers after requirements have been allocated and to capture cross-interface information in Interface Control Documents (ICD).

Understand the external and induced environments. Design teams must also understand the sensitivities of their system to nominal and off-nominal excursions in the external environment along with their associated probabilities of occurrence. The more complex the system the more difficult it is to understand the system sensitivities to the environment.

Controlling implications of new technology The mix of new and existing technologies in a design can add to complexity that affects safety and reliability. New technology can improve safety and reliability when carefully selected and applied, though new technologies often bring with them "unknown unknowns" that may represent safety and reliability risks.

In cases where new technologies are necessary, the systems engineer must help the design teams assess the maturity and identify potential interactions along with additional constraints and uncertainty the new technology might introduce. Introducing new technologies may make the system more reliable at maturity, but failures during the maturation process may make the system less reliable when considered over the life of the program. The systems engineer must understand how new technology introduces unknowns into the program, and what can be done to combat them. Examples include incorporating additional margin, extra testing, alternative flight manifests, and concepts of operations.

Applying Heritage and COTS system elements. Using "Heritage" and Commercial-Off-The-Shelf (COTS) system elements are often utilized as a way to reduce risk. However, COTS elements can introduce complexity and risk if they are not applied properly. COTS products bring with them design constraints, predefined interfaces, and operational constraints that the receiving system must accommodate. COTS elements, especially those with a proven flight track record, can improve safety and reliability, but it is their proper application and accommodation in a new and different application that represents a challenge to the systems designer. For COTS and heritage components, the design focus shifts from having to define the component's detailed requirements (as in newly developed items) to accommodating its constraints.

Ultimately, it is the responsibility of the engineering team to ensure that the benefits are realized and that unknowns are discovered before flight. The promise and advertised benefit of a new technology or COTS elements is often not realized in practice and should therefore be addressed from a risk perspective.

5. Conceiving the Right System (Section 2.3, Sections 4.0 through 12.0)

Conceive the right system conceptual design early in the life cycle by thoroughly exploring risks from the top down and using a risk-based design loop to iterate the operations concept, design, and requirements until the system meets mission objectives at minimum complexity and is achievable within constraints.

Defining the conceptual design early in the life cycle has a high degree of influence on the system's safety and reliability. Fundamental safety characteristics designed into the system will either put the program on a smooth trajectory, or one that will require numerous design fixes and operational workarounds to respond to emerging safety issues and failures. The objective of the early work should be to guide a balanced design by choosing either inherent reliability, like and/or diverse redundancy, or maintenance approaches to achieve safety and reliability.[4] Choices should be driven by mission objectives and constraints and justified by risk analysis. Choices then define the complexity of the system, how it can fail, how likely it could fail, what happens should it fail, and any coupling and cascading effects of the various system elements.

Since the early work has a high degree of influence on safety and reliability, it is critical that lower level requirements not be written too early unintentionally precluding other viable and safer solutions allowed by an alternate set of requirements.

In an ideal world, the design process is a series of iterative steps performed in a top down fashion along the life cycle guided by risk assessments that are sequenced so that uncertainty is progressively removed as design knowledge matures.

Below is one path applicable to the upfront work in the life cycle that can lead to such a design:

Step 1 Define needs, objectives, and constraints in clear and simple terms, and then capture them as the high level requirements. These form the validation basis for the subsequent work. The constraints can be visualized as the boundaries of a box as shown in Figure i-3. Subsequent steps seek to find solutions within box constraint boundaries and select one for development. Alternatives should be plotted against constraints forming a "surface" from which a solution is selected.

---

[4] A survey of existing proven designs and previously developed concepts can be useful at this point in the development, but constraining the process to only existing designs/concepts too early can lead to the "wrong design."

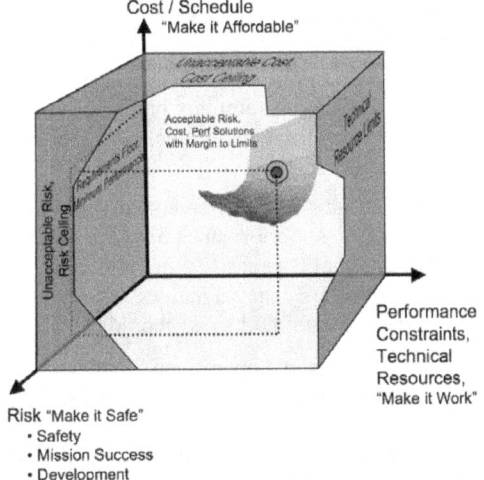

Figure i-3 Project Constraints Box Showing Alternatives as a Surface with Selected Solution

Step 2 Define the minimum set of functions necessary to accomplish the mission objectives.

- Identify and describe the functions the system must perform from a systematic top down perspective to fulfill mission needs and objectives.

- Clearly identify and distinguish functions necessary for safety and mission success. This distinction is critical for assessing and accepting a function's criticality, appropriate fault tolerance, and probability of failure.

- The identified critical functions should be used to set up the Product Breakdown Structure (PBS) (the source for a product-structured WBS) in a manner that prevents unnecessary splitting of safety critical functions that would complicate interface control and team understanding of adverse couplings.

- Defining the necessary functions allows a clear understanding or statement of the problem to help guide and define appropriate solutions. Often a solution becomes evident after a clear statement of the problem.

Step 3 Make it work. Create the simplest conceptual design of the contemplated system.

- Start with the simplest, most robust, and highest performance design option as the primary leg for accomplishing the mission functions identified above with inherent safety. The primary leg also forms the first leg when assessing fault tolerance. The simplest solution should lie within the constraint box boundaries with adequate margin for the succeeding steps below.

- If the simplest solution falls outside the constraint box, then there may not be a workable solution; start the process over again with an alternate set of needs, objectives, and constraints.
- If the solution falls inside the constraint box but is not viewed as viable or optimum, consider alternatives with different operational concepts, designs, or derived requirements as depicted in the iterative loop shown in Figure i-4.

Step 4 Make it safe. Add diverse or independent elements to the simple system of step 3 that operates at lower or even marginal performance but with higher reliability as necessary to meet safety needs. This additional leg adds to system fault tolerance, although it may be applied as the last leg not necessarily the second leg. A simple diverse system maximizes the independence from prime system faults and should be easier to understand and verify.

- Evaluate the conceptual design and operations concept to determine potential failure modes and safety impacts. Initially the evaluation must be performed from the top down starting from the mission level and consider each operational phase or operational system configuration of the mission. Utilize Functional Failure Mode and Effects Analysis (FMEA) (based on functions) and/or Fault Tree Analysis (top down based on undesired consequences) along with an integrating technique such as Event Sequence Diagrams (ESD) to identify Risk Drivers.
- Utilize risk and reliability modeling techniques to bound the likelihood of the identified safety drivers. Discussions and debates resulting from likelihood and consequence discussions are helpful for further understanding and exploring the risk drivers.
- Pay particular attention to common cause and common mode failures that may defeat the intended safety improvements of the additional elements.
- Iterate the candidate mission rules and procedures to safely achieve the minimum acceptable objective.
- Provide an abort mode for those phases of the mission where the likelihood or consequence of safety critical initiating events or consequences cannot be contained.
- An effective methodology is to start with the "end game" of returning the crew to Earth and continue to work backwards from Earth reentry to mission initiation assuring that safety and reliability are preserved during each operational phase of the mission. In other words utilize technical resources such as, mass, volume, power, etc, to get the crew home first.
- If the solution is not safe, consider alternatives with different operational concepts, designs, or derived requirements as depicted in the iterative loop shown in Figure i-4.

Step 5 Make it reliable. Consider additional elements or other "legs," preferentially an additional primary leg of equivalent performance but not necessarily identical design for mission success. Additional legs for mission continuance add to system fault tolerance. Determine if the addition of the mission success leg leads to a safer system by considering all the potential dependencies.

- Utilize risk and reliability modeling techniques to estimate the effects of one alternative over another. If an alternative reduces overall risk is affordable, add it; if not, be sure that the implications of accepting this risk are understood. Again, discussions and debates resulting from likelihood and consequence discussions are helpful for further understanding and exploring the risk drivers.
- Pay particular attention to common cause failures that may defeat the intended safety and reliability improvements of additional elements. Strive for designs that will limit the occurrence or consequence of common cause failures.
- Consider the maturity and complexity of the system when addressing how to mitigate "unknown unknowns." This may drive additional features to facilitate testing and verification needs; for example, additional test points or data recorders.
- If the solution is not reliable, consider alternatives with different operational concepts, designs, or derived requirements as depicted in the iterative loop shown in Figure i-4.

Step 6 Make it Affordable. Estimate cost and schedule to develop, produce, and operate the system design of steps 2 through 4.

- Upfront design work has a high degree of leverage on the system's cost since these early activities expend around 10 to 15 percent of the project cost yet commit in excess of 50 percent of the total run out costs.
- Iterate the operations concept, the design, or the derived requirements as necessary to satisfy constraints, go back through steps 2, 3, 4, and/or 5 as necessary.

Step 7 Capture the Conceptual Design

- Capture the decisions and supporting rationale of steps 2, 3, 4, and 5 as the derived requirements, baseline operations concept, and baseline conceptual design.
- Consider all the legs of the system design when assessing system fault tolerance utilizing the rationale developed in the above steps to justify any differences between the selected approach and the starting point of two-fault tolerance.
- Develop a program plan that tentatively defines prioritized requirements for each system element in the PBS, allocates physical and resource constraints to each, describes a system acquisition strategy, and assigns management responsibility for each effort. At the completion of this step, a safe and reliable system, producible at minimum cost, schedule, and complexity has been identified.

The previous steps describe a methodology for developing a design that can meet the needs and operational requirements through an iterative loop until performance, cost/schedule, and risk constraints are met. Even though the process is described in a step-by-step fashion, aspects of performance, safety and reliability, and affordability are not independent quantities and should not be considered independently in the process of design. While safety is of paramount importance, the implied order or hierarchy to the design process obligates the designer to make the design work first, make it safe and reliable, and then assure it is affordable. This is because affordability is moot if the design will not achieve a reasonable level of safety and reliability, and safety and reliability are moot if the design does not function.

Utilizing this type of integrated methodology described above and shown in Figure i-4 provides affirmative rationale for the system design, its complexity, and the existence of each system element. This approach may lessen the likelihood of having to lop off pieces of a design to get it back "in the box." Lopping invariably leaves the system in a less cohesive state, vulnerable to unexpected interactions and other shortfalls.

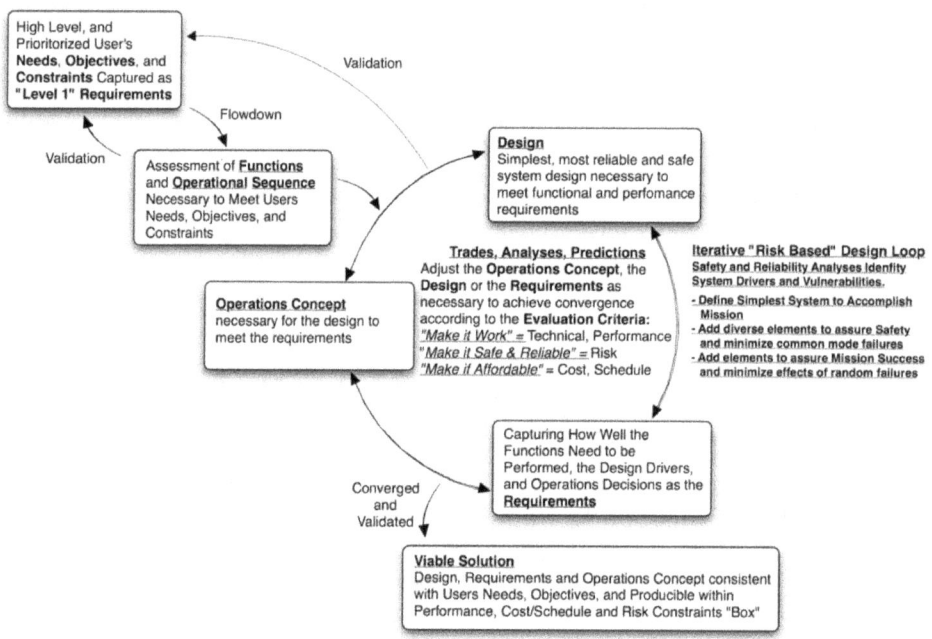

Figure i-4 Objectives Driven and Risk Based Iterative Design Loop

- Needs and Objectives drive the identification of functions
- Functions drive an iterative risk based loop that converges the operations concept, the design, and the requirements towards a viable solution
- The viable solution meets users needs and objectives within program constraints

At any point along the design life cycle teams can ask; "Is this the correct solution, and why is it safe?" One effective technique is to start with the integrated system block diagram and determine if sufficient justification exists for the existence and configuration of every system element on the block diagram with rationale grounded in safety and mission success. If team members are not sure that the solution is the correct solution then retracing the steps described above can help with providing the desired justification and rationale for the design choices.

6. Building the System Right (Section 2.4, Sections 4.0 through 12.0)

Build the system right by applying a multilayered, defense in-depth approach of following proven design and manufacturing practices, holding independent reviews and inspecting, and employing a "test like you fly, fly like you test" approach to assure that the system is safe and reliable.

Building a safe and reliable system requires the team to follow sound design and implementation processes with emphasis on safety and reliability during each life cycle phase. After the architecture is defined, the selected concept must be matured into hardware, software, logistics, etc., through the "Preliminary and Detailed Design," "Manufacturing and Assembly," "System Integration and Verification," and "Operations" life cycle processes as shown in Figure i-7.

Assuring safety and reliability in a complex system operating in uncertain environments argues for overlapping and diverse methods to not only develop the system properly but also to provide maximum coverage for identifying and screening potential problems. Deploying a multilayered approach as shown in Figure i-5 not only synergistically drives the system to function as intended and required, but also allows the discovery of potential catastrophic problems. Diverse layers depicted as multiple nets in Figure i-5 include:

- **Sound design and manufacturing processes** based on proven practices defining the intent for producing a safe and reliable system. Performance predictions and analytical assumptions are validated by test verified models or test of engineering and/or qualification hardware. Well-controlled manufacturing process, including statistical controls where appropriate, maintains or enhances quality and reliability.

- **Independent reviews by experienced peers** allow the team to benefit from the experience of others not involved in the program. Reviews should be set up to benefit the program team and should be viewed as a welcome input to system safety and reliability. Review teams should not only utilize their experience to critique but should also suggest potential solutions.

- **Inspections and walk downs** by not only Safety and Mission Assurance (S&MA), but also Engineering and the Crew to assure the system is produced as intended.

- **"Test Like You Fly"** approach broad enough to include not only requirements verification but also validation of the system through performance, mission simulations, end to end tests, and joint integrated simulations including off-nominal scenarios that assure the system can in fact accomplish the mission with the intended safety controls and mission success robustness. The "Test Like You Fly" approach includes critical examination where such testing is not possible or accomplished in pieces to assure sufficient test coverage over the expected flight environments and operational sequences. Testing provides the last line of defense and opportunity to discover unexpected interactions and the ability to validate and verify models used during design. A difficult challenge of discovering unknowns lies in assuring that complex systems are testable and verifiable prior to flight. These test approaches therefore need to be factored into a testable design.

- **"Fly Like You Test"** operations approach to limit the chance of encountering an unexpected interaction among system elements and their environments not previously explored during test and verification

No single layer will function perfectly in producing a safe and reliable system. All layers are needed to provide the opportunity and coverage necessary to assure that the details have been adequately considered. Proper application of process at each layer provides the affirmative body of evidence that the system is safe and ready to fly.

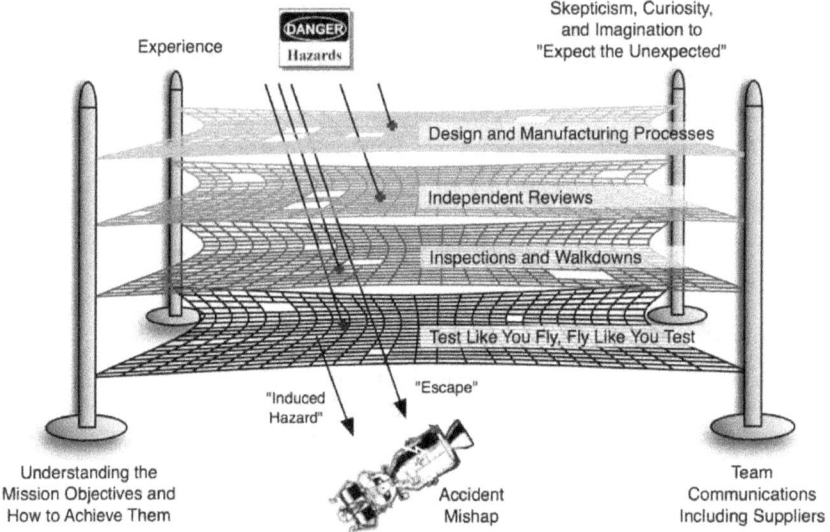

Figure i-5 Multilayered Approach to Produce a Safe and Reliable System and Screen for Potential Problems

- **Dense and diverse nets with solid supporting poles serve as barriers or screens preventing hazards from causing accidents or mishaps**

- **Multiple imperfections in the nets or supporting poles may allow hazards to result in accidents or mishaps**
- **Avoid inducing hazards or latent failures into sensitive system elements by processing steps**

The poles that support the nets in Figure i-5 are critical to their integrity and effectiveness. Development of a safe and reliable system requires access to experience, team members who understand the mission needs and objectives, have curiosity, skepticism, and imagination to identify what can go wrong, and can assure good communications among team members and suppliers. It is the role of program management to assure that the objectives are communicated, appropriate experience is applied, and that the team is encouraged to communicate safety and reliability issues.

7. *Integrating Risks Throughout the Life Cycle* (Sections 2.5 and 3.0)

Seek and collect warning signs and precursors to safety, mission success, and development risks throughout the life cycle, and integrate those into a total risk picture with appropriate mitigation activities.

An integrated risk management assessment process throughout all phases of the system's life cycle is essential to achieving a safe and reliable system. Early identification and resolution of potential problems is key to effective application of technical, cost, and schedule resources. Each layer within the multilayered approach provides a mechanism for identifying and collecting warning signs and precursors to conditions that could impact safety and reliability. Tracking and trending these warning signs throughout the life cycle is key to acknowledging and quantifying the risk. Team members must pay particular attention to these warning signs and affirmatively resolve them with rationale describing why the system is safe.

Sometimes risk is accepted and accumulated in small increments. Each of the individual risk increments by itself may not appreciably increase total risk. However, a large number of small risks can aggregate, couple, amplify, and combine to a much higher risk state. Risk management activities should recognize this effect and provide a mechanism to inform decision makers of the total aggregate risk.

Distinguishing risks by their "safety," "mission success," and "development/programmatic" consequence types encourages the team to discuss and focus on what is at stake. Decision makers can then integrate a total project risk picture and can decide where to apply resources. Distinguishing among these risks types can help teams make difficult choices when evaluating safety versus mission success versus development risks.

A total risk picture allows informed deployment of technical, cost, and schedule resources to obviate or mitigate risks. Figure i-6 shows how risk information flows from different sources and is evaluated in total.

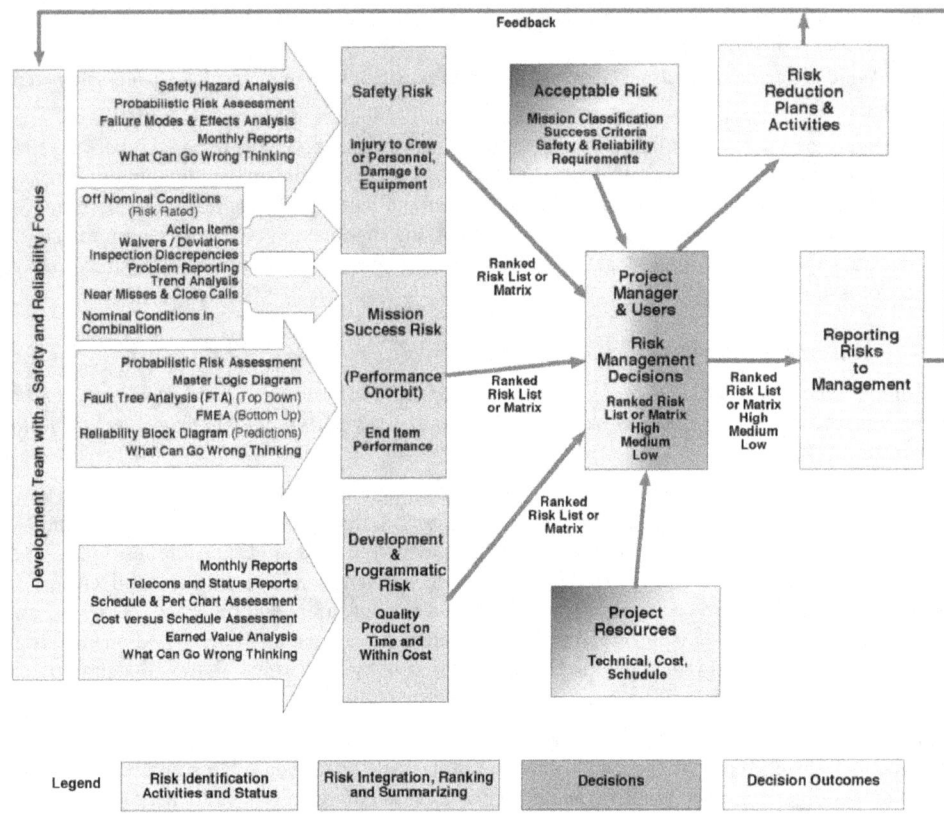

Figure i-6 Risk Information Flow

- Team identifies risk by consequence using targeted methods
- Risk management boards integrate and rank risks
- Decision makers apply technical, cost, and schedule resources to maintain an acceptable level of risk
- Risk reduction plans and activities feedback to development and operational activities
- Risks reported to management and team members

Some of the elements critical to producing safe and reliable systems require the following emphasis in risk management:

- Integrate a total risk picture from each contributory increment of risk.
- Provide a method of distinguishing, comparing, evaluating, and prioritizing risks with differing and sometimes competing consequences to allow informed application of technical, cost, and schedule resources to obviate or mitigate risks.
- Foster curiosity, skepticism, and imagination to identify risks, seek warning signs, and precursors to potential failures and problems. Encourage an integrated team of designers, operators, crew, safety, and reliability analysis personnel to assess risks from balanced optimistic and skeptical viewpoints.
- Start risk assessments early in the life cycle in a top down fashion to provide the highest opportunity to obviate and mitigate risk while a wide range of opportunity exists at low cost. Remember and revisit the top down perspective later on the life cycle.
- Validate system risk assessments with historical knowledge of failures and their causes in comparable systems.
- Identify residual risks, close calls, anomalies, or issues that do not have a definitive cause or corrective action, and where their reoccurrence can result in significant safety and mission success consequences. Utilize affirmative rationale and data grounded in the scientific method to support flight readiness assessments.

Conclusions

The guiding principles, applied at the system and subsystem level along the life cycle, seek a design that is safe and reliable. A necessary ingredient will be the leadership to ensure that safety, reliability, robustness, and fault tolerance are incorporated into the design and the end item at all levels. Integrating the various system elements into a cohesive whole and assuring the integrated system is safe remains a prime challenge to the engineering and management teams. Therefore, the resulting system must be a combined effort of systems, subsystems, and component engineering, manufacturing, S&MA, test, and operations teams that execute the "right" processes at the "right" time throughout the system life cycle. The team should be led with a management philosophy that fosters a safety focus along the life cycle and encourages open communications for surfacing and resolving safety concerns with affirmative evidence that the system is safe to fly.

Figure i-7 shows one way the methodologies discussed in this executive summary integrate into an overall view of what is done to ensure a safe and reliable system. The safety focus along the life cycle and team forms the foundation. The seven principles described above rest on this foundation and synergistically guide the development of a safe and reliable system.

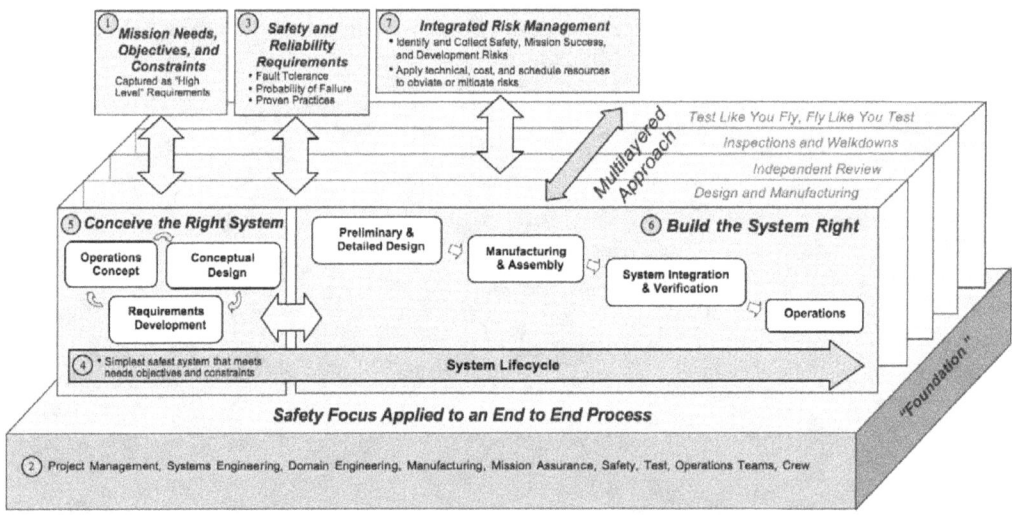

Figure i-7 Guiding Principles Applied to End-to-End Development of a Safe and Reliable System

1.0 Introduction

NASA is in the process of designing the next generation crewed exploration-type space flight missions. The previous generation, the Space Shuttle, was designed more than 30 years ago, with a different set of mission/functional requirements. In fact, this new design is expected to be far more like the first generation Mercury/Gemini/Apollo Missions than that of the Shuttle. The optimum methodology for specifying requirements for mission success and crew safety are now under consideration.

What factors should be considered in establishing such requirements? This question prompted the Astronaut Office at NASA JSC to commission this report under the auspices of the NESC. A team was chartered by the NESC Systems Engineering Office (SEO) (refer to Appendix B) to assess best practices for developing a crewed space vehicle that ensures safety of the crew and is both reliable (free of failures throughout its mission) and robust (tolerant of unexpected conditions should they arise). The existing NESC discipline Super Problem Resolution Teams (SPRT) were leveraged extensively to capture agency knowledge and best practices, particularly in systems and processes that drive system safety and reliability. Outside organizations, such as The Aerospace Corporation, Draper Laboratory, Valador, and Bay Engineering, were utilized in the gathering of relevant historical systems data, best practices in design, verification, and assessment of reliability. Copious references are included, which benchmark documents across the aerospace industry, both inside and outside NASA, for the further insight.

To cover the wide range of considerations of designing for safety and mission success, this report is divided into the major sections listed below, along with questions that were asked of the NESC by the astronaut office.

Section 1.0 – Historical Perspective

- Which systems have been reliable?
- Which systems have required redundancy?
- How did the reliability mature for Apollo, Space Shuttle Program (SSP), and ISS within each system?

Historical information can provide helpful guidance on how to make future spacecraft as successful as possible. This section will provide an overview of how missions including Apollo, Shuttle, and ISS, have handled safety and reliability and provide an examination of known causes of space mission failures.

Section 2.0 – System Engineering with a Safety and Reliability Focus

Section addresses the following questions:

- What methods are there for designing-in and assuring reliability of critical Earth-to-orbit systems?
- How is it known where redundancy is and is not required?

- Which types of systems are highly reliable without redundancy and which systems require redundancy in order to achieve high reliability?
- How are the reliability requirements specified?

Achieving a safe and reliable system requires an intently focused effort at every phase of the life cycle. This approach entails following a sound design process with emphasis on safety and reliability drivers at each life cycle phase. The process must be based on the foundational principles of safety and reliability. Elements of the process are illustrated in Figure i-1. Section 2.0 will address Systems Engineering (SE) technical considerations and methods for apportionment of project-level resources throughout the design process flow. These sections seek to describe methodologies for how to design mission success into a program from the ground up. Where and why redundancy is applied will be discussed with the recognition that mission success is far more complex than the simple application of redundancy. Section 2.0 subsections are described herein.

- Section 2.1, "The Right Work at the Right Time with the Right Teams:" Before even getting started, System Engineers must stress the importance of setting simple and concise objectives, following the right process, and performing the right functions when they are needed. Results of trades performed early in the life cycle have a significant influence on systems safety and reliability as well as costs. None of this can be achieved without the right teams that are properly integrated. Getting the right teams in place is critical to the entire SE process.
- Section 2.2, "Defining the Requirements for a Safe Human Rated System:" The overall objective for specifying requirements for human rated systems is to provide guidance for design and a basis to verify that the system, once produced, is safe. The integrated teams must understand this philosophy from the beginning, as this requirement will effect all aspects of the system including: redundancy schemes, backup systems, abort systems, and operational procedures.
- Section 2.3, "Conceiving the Right System, Critical Activities Early in the Life Cycle:" There are key activities that must be performed early in the design process. A risk-based system design loop is a key component in this early development as this loop attempts to converge on a set of self-consistent requirements, design, and operations concept. The requirements set the stage for the design and are based on the mission objectives. This loop also helps to put emphasis on the design drivers and strives to maintain the simplest design architecture to avoid unnecessary complexity that increases risk.
- Section 2.4, "Implementing the System Right, Achieving a Reliable System:" After the architecture has been defined, the concept selected must be matured into hardware, software, logistics, etc. through the design process. This is a major activity that is based on classical SE practices with a focus on technical integration of system, subsystem, and component designs. The system, subsystems, and components are integrated by SE using validation, trades, risks and sensitivity studies. In addition to providing a brief overview

of the design process, this section will discuss a multilayered approach for developing a safe and reliable system.

- Section 2.5, "Integrating Risk:" An integrated risk management process must pervade all phases of the system's life cycle. This process provides Program management with information necessary to identify, understand, and evaluate risk trades throughout the life of the Program. This section provides a systems perspective of integrating risks from multiple sources utilizing different methods to obviate or mitigate the highest risk when considering technical, cost, and schedule.

Section 3.0 – Safety and Reliability Analysis throughout the Life Cycle

This section will address the role of safety and reliability analysis throughout the design process flow from Pre-Phase A to Phase E. There will be a focus on reliability assessments during each phase, and what inputs are needed to perform a proper reliability assessment. In addition to discussion about performing the reliability assessments, details are provided on how to use reliability assessments during each phase, and how these assessments and analyses evolve throughout the design process.

Sections 4.0 - 12.0 – Domain Engineering with a Safety and Reliability Focus

Discipline areas provide perspective on those aspects of the DDT&E process that are most critical or unique to their part of the system to ensure safe and reliable design, based on the extensive experience of team members, accepted industry practice (including standards), and lessons learned from preceding missions. Discipline sections are organized to correlate with the SE perspective identified in Section 2.0, as follows:

1. Interfaces within and outside their subsystem
2. History relevant to reliability and robustness
3. Architecture
    a. Trade studies, along with evaluation criteria necessary to converge design, operations concept, and derived requirements
    b. Reliability drivers
    c. Technical integration at system level
4. Design, Build, and Operations
    a. Best practices

Also included in each of the sections is a list of "indicators," factors by which an observer can judge whether a design is reliable and robust.

In summary, this report gathers information necessary to assess and document the prevailing methodologies for maximizing and assessing critical spacecraft systems safety and reliability.

This report is not intended to duplicate or update existing references and is not intended to replace existing standards and policy.

1.0.1 Definitions

Cascade Failures: Multiple failures that occur because the consequence of one failure results and cascades into the failure of another system element.

Common Cause Failure: The failure of multiple items occurring from a single cause that is common to all of them.

Common Mode Failure: The failure of multiple identical items that fail in the same mode. Note that common mode failures are a particular case of common cause failures.

Dependent Failure: A failure whose likelihood is not random and dependent upon another system element or environment. A dependent failure is a failure of a component or system that is not statistically independent of another failure. That is, its probability of failure is different if another component has failed.

Failure: A required function or specified service of a system, device, software, or system operator ceases

Fault: An incorrect state of hardware or software resulting from failures of components, functional upsets, operator error, or incorrect design. Faults include failures of system elements.

Fault/Failure Tolerance: The ability to operate/survive in the presence of faults or failures

Fault/Failure Avoidance: The potential for faults/failures are obviated or mitigated through screening processes.

Independent Failure: A failure whose likelihood is random and not dependent upon another system element or environment. An independent failure is a failure of a component or system that is statistically independent of other failures.

Malfunction: A failure is a cessation of operation; in a malfunction, the system continues to operate but in an unexpected manner.

Mission Success: Accomplishment of mission objectives and meeting mission requirements

Quality: "The totality of features and characteristics of a product or service that bear on its ability to satisfy stated or implied needs." [ISO8402] Quality is characteristic that a component or system possesses at a fixed point in time, usually at the beginning of life.

Reliability: The ability for the system to continue to perform its function over its mission time. Reliability is introduced into the design by the establishment of margin against the operational environment. While quality is necessary it is not sufficient to indicate that, the unit will function over time in its operational use and environment. "The ability of an item to perform a required function, under given environmental and operational conditions and for a stated period of time." [ISO8402]

Robustness: The ability to operate/survive in the presence of an abnormal or unexpected environment during operation. Robustness is related to the margin against system limits and its insensitivity to changes in the environment and operational sequence.

Safety: Protecting against consequences of failure, damage, error, accidents, or any events that could be considered dangerous to human life. Protection can be applied to either the cause or the consequence of something that is not safe.

1.1 Historical Context of Space Systems Safety and Reliability

From its inception in the late 1950s, the history of spaceflight, particularly human spaceflight, has been distinguished by engineers accepting complex and demanding challenges to the edge of imagination and then meeting those challenges. Early on, developing and operating systems to venture into unknown territory beyond the friendly confines of Earth has challenged the best minds that the United States had to offer, as well as the utmost in government and industrial capability. Such efforts achieved a high level of success, with but a few unfortunate instances of failure.

In the case of the Apollo 1 fire no one had "imagined," as Frank Borman noted, the hazard of common materials in a high-pressure oxygen environment resulted in a highly flammable and dangerous situation for that ground test configuration. Apollo 13 on the other hand demonstrated how a ground test set up could compromise the reliability of the flight system, and that flexibility, robustness, human ingenuity, and creativity can overcome serious obstacles to safely return the crew to Earth.

During the mid-1980s through the 1990s, a disturbing trend of increasing robotic spacecraft failures and two highly visible Shuttle failures occurred. This report attempts to convey or reinforce the keystones of mission success, in safety and reliability terms, to the designers and operators of the new generation of crewed space systems.

A brief look back at the history of spaceflight and how safety and reliability approaches evolved will help determine where to go from here. Early in space exploration there was limited knowledge and confidence in space systems hardware performance in the space environment. At the time of Alan Shepard's Freedom 7 mission in 1961, powered air travel of any kind was less than half a century old. Robert Goddard had launched the first liquid fueled rocket only 35 years prior to Shepard's flight! Large booster rockets had relatively low performance and were not entirely reliable. Complex electronics were new and individually unreliable. Astronauts were exclusively military test pilots flying high-risk missions. However, protecting their lives was of paramount concern.

NASA set about a program to develop missions with high system reliability. Methods were continually being devised to recover from potential failure or preclude it altogether. Since the ability to predict performance was rudimentary, equipment was over designed with high margins. System reliability was achieved through extensive test programs and judicious use of redundant components, crew intervention, or ground command to recover from a box failure. As Apollo 13 showed, however, malfunction is not always due to random box failure that could be mitigated by redundancy. The application of diverse systems or the alternate use of existing systems at the sacrifice of performance, such as a Launch Escape System, should also be considered by design rather than relying on happenstance and ground crew resourcefulness.

As flights increased, our understanding of vehicle behavior increased. With understanding, thanks in large part to the advent of large-scale digital computers and the software to command them, the ability to accurately predict performance increased, allowing design margin requirements to decrease. Tests that may once have been considered crucial verifications, but having shown few failures lately, were omitted.[5,6] In some cases, where knowledge and experience had been accumulated, attempts to control costs resulted in the omission of specifications and standards,. Cost and schedule pressures also resulted in less focus on SE with sometimes significant adverse consequences.

As this section has described, safety and reliability approaches have evolved in the past 35 years. Table 1.1-1 summarizes how each of the major human spaceflight systems, robotic spacecraft, and aircraft has generally addressed reliability requirements and implementation approaches. As evidenced in the table, differences in approach to the implementation of crew safety and reliability requirements exist across programs. This is because fault tolerance is a functional requirement applied at the mission level, not a hardware requirement. That is, for any given function the mission must perform, the prescribed number of faults (currently two for crewed missions) must not result in loss of crew, but may result in reduced performance. There are a number of possible techniques by which fault tolerance requirements can be satisfied, using redundancy and/or diversity, or waived by employing ultra-highly reliable components or systems. Which techniques are selected for any given mission is a function of a number of factors, including constraints (size, weight, power, and cost), technology readiness level (TRL) of hardware, predicted reliability, crew in-the-loop architecture, etc. Sections 2.0 and 3.0 will address how these factors influence mission acquisition, including reliability requirements.

Section 2.2 discusses potential reliability requirements. The discussions are based in the various approaches adopted by the programs listed in Table 1.1-1. How these projects specified their reliability requirements influenced the recommendations in Section 2.2.

---

[5] Tosney, W., Faster, Better, Cheaper, The Aerospace Corporation, Nov 8 2000
[6] Tosney, W., Quintero, A, Orbital Experience from an Integration and Test Perspective, The Aerospace Corporation, Journal of the IEST, Nov/Dec 1988

Table 1.1-1 Comparison of Historical Reliability Approaches

Mission	Description	Safety & Reliability Philosophy, Mission Life	Techniques for Reliability, Robustness, Redundancy, and Diversity
Mercury	Sub-orbital proof of concept and Single Crewman Orbital Operations	No Single Failure Results in Loss of Mission. No Single Failure during an abort results in loss of crew (Ref 1) Short Mission Life, Hours to Days.	Launch Escape System, Blunt Entry Body, Single Electronics Unit with Crew based diverse manual backup systems. Ground Support from multiple distributed ground stations / control centers to guide operations and support anomaly resolution.
Gemini	LEO, Proof of concept Orbit/Return, Docking in Preparation for Lunar Landing Missions	Probability Mission Success .95 Probability Safe Crew Return .995 (Ref 2)	Launch Escape via Ejection Seats, Blunt Entry Body, Single Electronics Unit with Crew Based diverse manual backup systems. Redundancy (functional & like) for all systems effecting Crew Safety. Mission Life Days to 2 weeks. Ground Support from centralized control centers to guide operations and support anomaly resolution.
Apollo	Moon Rendezvous/Landing/Return	Probability Mission Success .90 Probability Safe Crew Return .999 (Ref 3)	Launch Escape System, Blunt Entry Body, Single Electronics Unit with Crew based diverse manual backup systems, 3 Fuel Cells and power buses. Mission Life 1 to 2 week Ground Support from centralized control centers to guide operations and support anomaly resolution.
Skylab	LEO Long-term Workshop		Dual Redundant Computers, Repair via Crew Repair, EVA. Mission Life years. Ground Support from centralized control centers to guide operations and support anomaly resolution.
Shuttle	Space Access, Launch and Return (sat deploy)/ LEO Experiment platform/ Aircraft-type landing return	System: Fail Safe including aborts Avionics: Fail Operational, Fail Safe, Two Fault Tolerant. Exceptions Managed via Critical Items List (CIL) Retention Rationale	1st Stage Launch Escape Not possible, First Flights provided Ejection Seats, Return to Launch Site and Trans Atlantic Abort Landing site. Manual Diverse Backup to Landing not possible, Triple and Quad Redundancy, 5th Identical Computer contains diverse backup software. Mission Life 1 to 2 weeks, Vehicle Reuse, Repair via Crew Repair, EVA. Ground Support from centralized control centers to guide operations and support anomaly resolution.
ISS	LEO Long-term Experiment platform On-orbit assembly proof of concept	Two Fault Tolerant, Fail Operational, Fail Safe, or Exceptions Managed via Design for Minimum Risk	Redundancy Mission Life Decades, Repair via Crew Repair, EVA. Maintenance and Repair Supplies delivered via Shuttle. (Shuttle availability could effect ISS reliability) Ground Support from two centralized control centers to guide operations and support anomaly resolution.
Robotic	Various LEO, GEO & Interplanetary	Varies, Often "No single Point Failures" with Exceptions managed	High Margins in design. Often a diverse reduced function Safe Hold Mode providing the Ground Time and Opportunity to resolve problems. Increased onboard autonomy for interplanetary missions. Rarely Beyond Dual Redundant. Onboard autonomy for normal mission, often some time critical onboard fault detection and correction Long mission life typically >2 years up to 15 years continuous Ground Control from mission specific control centers critical for resolving anomalies.
Commercial Aircraft	Transport Public	Fail Safe, 1) No single failure shall prevent continued safe flight and landing. 2) System designed such that the probability of failures in combination must not exceed 1x10-9 per flight hour 3) Follow established principles (Ref 4)	Various Triple or Quad Redundant systems either with or without diverse manual / mechanical backup. More recently without diverse backup
Military Aircraft	Transport Military hardware, personnel, and deliver weapons	F-16 Requirement = 5E-05 / Hr Observed Failure Rates / Hr (Ref 5) F16 3.9E-05	Various Triple or Quad Redundant systems either with or without diverse manual / mechanical backup. More recently without diverse backup

Ref 1 French, John C. Mercury Project Summary SP-45, Chapter 6 Reliability and Flight Safety
Ref 2 Gemini Mid Program Conference, Spacecraft Reliability and Qualification, Chapter 10, 2/23/1966
Ref 3 Hershkowitz, B.H, Apollo Technical Manual, Reliability, ARM-10, 10/15/1963
Ref 4 FAA 25.1309, Advisory Circular 25.1309-1A
Ref 5 F-16 Systems Safety Handbook, 16PR361 http://www.f-16.net/varia_article5.html

1.2 Examination of Failure History Data

Space missions are among the most difficult endeavors undertaken by humans. Despite the best efforts of a government/industry partnership comprised of a skilled and dedicated workforce, there are occasionally failures. Extensive anomaly data was evaluated to assess the origin of space vehicle failures. Along with data from The Aerospace Corporation Space Systems Engineering Database and Rome Air Development Center (1985), failure investigation documentation was also examined for guidance, including:

- The President's Commission of the Challenger Accident (Rogers Commission)
- The Columbia Accident Investigation (CAIB) Report
- The Report of the Defense Science Board/Air Force Scientific Advisory Board Joint Task Force on Acquisition of National Security Space Programs (Young Commission)

Another valuable resource in understanding space mission failures is The Aerospace Corporation's *100 Questions for Technical Review*. In this document, the author has described, in a consistent summary method, catastrophic anomalies that have resulted in mission loss along with their established root cause. That data is used to develop questions that a prospective project reviewer should ask to determine whether a mission is likely to be prone to repeating the same mistake.

It should be noted that correlation to an overriding specific deficiency is not always obvious from the data and there is a concern that the relative infrequency of failures may not allow for a statistically valid data set. However, there are trends that can be inferred from the available data. The following paragraphs will identify and interpret significant trends from past space vehicle failures. A summary list of observations follows.

- Electrical, electronic, and electromechanical (EEE) parts failure rates are decreasing.
- Proper application of EEE parts in multi chip modules (MCM), hybrids and printed circuit boards remain a challenge.
- No single subsystem is immune from failure.
- Proper testing is key to mission success.
- Failures are likely if initial cost and schedule allocation are not scoped appropriately for the project's complexity.
- Anomalies must be thoroughly understood.
- The lack of proper standards, contractor oversight, and test will cause an increase in failures.
- Software Failure Rates are increasing.

- Design, Environment, Parts, and Workmanship account for the majority of the failure causes seen today.

1. Mission Phase and Risk: Figure 1.2-1 shows all known human spaceflight failures, both catastrophic and significant, as a function of relative mission timeline. Figure 1.2-1 shows that failures can occur at any time and significant failures have occurred during on-orbit operations. It can be inferred that space vehicle robustness precluded a number of significant failures from becoming catastrophic. Unlike ascent and entry, the crew has time to evaluate, troubleshoot, and respond to on-orbit failures (i.e., Apollo 13). One notable statistic in Figure 1.2-1 is that loss of crew has only occurred during ascent and re-entry, the two most dynamic phases of a mission. During high risk, highly dynamic phases of missions it is important to include response time and automation into redundancy and backup reliability trades. Sections 2.0 and 3.0 will discuss the importance of considering the different aspects involved in all mission phases when selecting architectures and when deciding where to buy down risk.

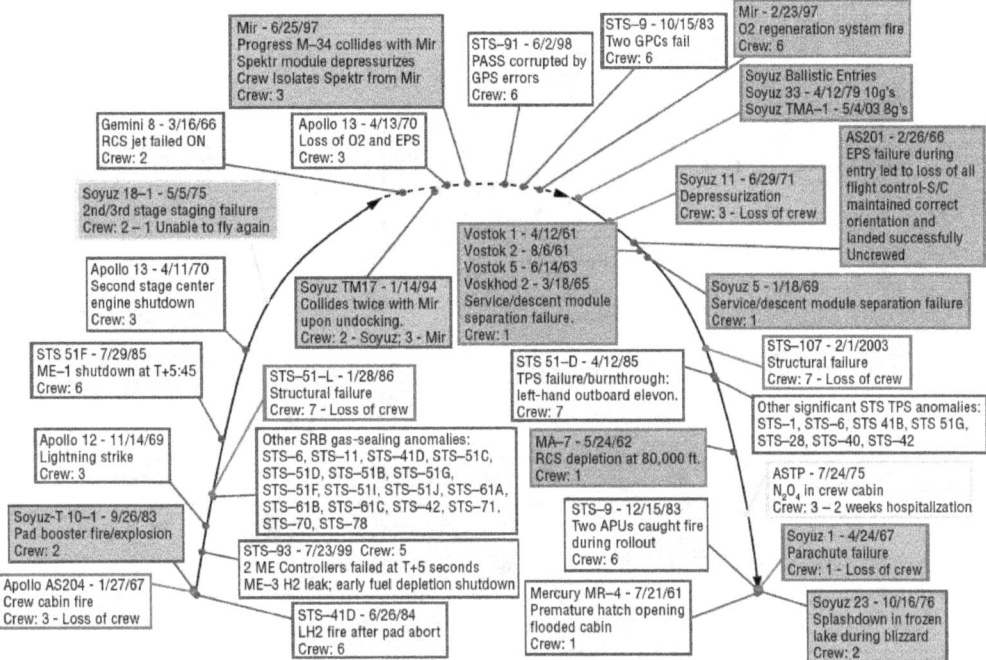

Figure 1.2-1 Significant Human Space Vehicle Failures
Ref: OSP-ELV Human Flight Safety Certification Study Report
Legend: Red Outline Box = Loss of Crew
Yellow or Orange = Crew Health Threatened
Green Outline = Significant Event / Close Call, Crew Unaffected

2. EEE Part Failure Rates: An area with a clear trend in failure data examined is electronic piece parts. As Figure 1.2-2 shows, individual parts failure rates have undergone a significant reduction over time due to improved manufacturing techniques. However, the investigation also found that, when individual piece parts are aggregated into higher-level components, such as hybrids, MCMs, and other sealed cavity assemblies, reliability related problems are disturbingly high. Recent examples include hybrid DC-DC converters and solid-state switches.

Anecdotal evidence indicates that the use of physics-based modeling and other modern computer aided design tools allows application of components closer to the manufacturers' published limits. This reduces the margin available to accommodate off-nominal conditions and statistical variation of components. This places renewed emphasis on the proper application and derating of EEE parts. Highly accelerated life testing (HALT) can be used to increase robustness; particularly for electronics parts and assemblies exposed to harsh environments.

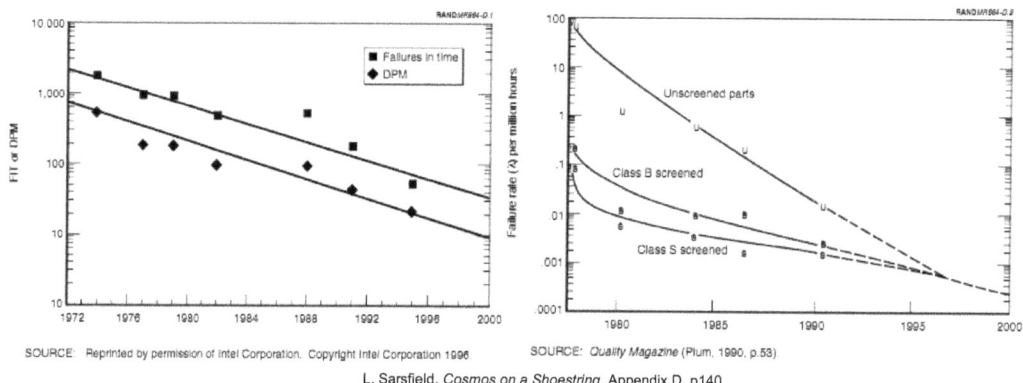

L. Sarsfield, *Cosmos on a Shoestring*, Appendix D, p140

Figure 1.2-2 Individual Parts Failure Rates

3. Adequacy of Testing: Spaceflight missions rely heavily on testing to uncover defects in either design or workmanship prior to launch. Mission success is directly related to the rigor of its test program, as shown in Figure 1.2-3. The figure shows Class A missions that follow rigorous testing to MIL-STD-1540 have a lower failure rate than programs with less rigorous testing. A test program that adheres to the principles, if not the letter, of MIL-STD-1540 is far more likely to be successful than one that does not. Section 2.4 provides further explanation on the need for incorporating full test programs.

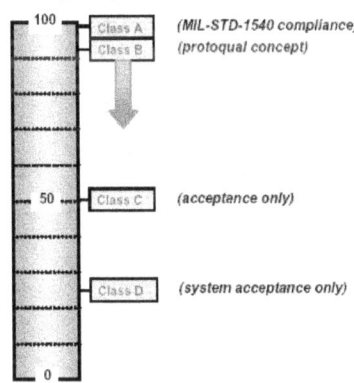
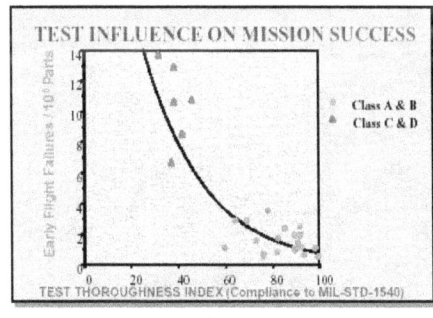

- On-orbit mission degrading failure (MDF) rate can be correlated per 100k piece parts
- Computation provides a measure of robustness of the environmental testing program used to approximate risk and highlight areas of concern

**Figure 1.2-3 Test Assessment and Risk Management
Aerospace Institute Course, W. H. Tosney, 2005**

4. Complexity of Systems:

Many believe that there is a critical level of complexity, which will almost assure system failure since, at some point; even the best Program Manager will not be able to keep track of the true Program status. Critical flaws will find their way into either the design or operating procedures.

Studies clearly indicate that more complex missions that have not been allocated additional acquisition cost and schedule based on complexity, have a higher likelihood of failure as presented in Figure 1.2-4. There are two ways to interpret the graphs. One interpretation is that a complex mission must be provided adequate budget and sufficient time to be successful. Some, however, would argue that this is not what the data actually shows. Another interpretation suggests that if a mission is sufficiently complex, it will probably fail. While it is unlikely that either of these extreme positions is always correct, the data clearly shows a strong correlation between mission complexity and mission risk. Thus, it is prudent to limit complexity to the minimum required to accomplish the mission objective.

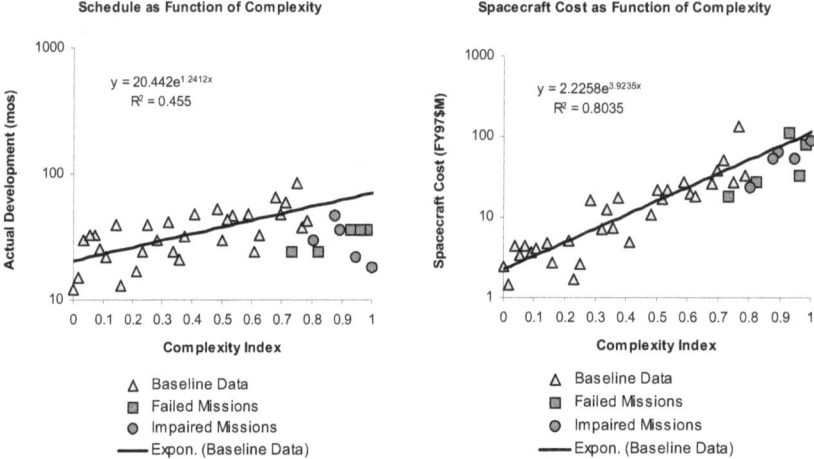

Figure 1.2-4 A Complexity-Based Risk Assessment of Low-Cost Planetary Missions: When is a Mission Too Fast and Too Cheap (David A. Bearden, Fourth IAA International Conference on Low-Cost Planetary Missions, JHU/APL, Laurel, MD, May 2-5, 2000)

5. Reliance on Past Successes: Rather than the tracing of each flight anomaly to root cause and objectively assessing its risk to mission success, an outlook has developed whereby past performance is thought to be indicative of future success. This is clearly not the case. Anomalies must be thoroughly understood in the context of mission success and crew safety. Corrective action should be taken to ensure that those so indicated are rectified such that they do not reoccur. In the cases where identifying a definitive cause or implementing a definitive corrective action is not possible, affirmative evidence grounded in the scientific method must be supplied to justify the system is safe to fly. Both the Rodgers Commission Report on Challenger and the CAIB Report on Columbia caution against relying on apparent past success as an indication of safety especially in areas where off-nominal conditions that threatened safety did not initially result is a serious problem.

6. Reduced Visibility during Acquisition: In the 1990s, the industry underwent a major modification in prevailing acquisition philosophy, called "Faster, Better, Cheaper" in NASA and "Acquisition Reform" in the Department of Defense (DoD). In these approaches, limits were placed on cost and schedule, while performance expectations were kept the same or increased. To achieve those goals, the requirements process was reduced, military and government standards were no longer levied on contracts, testing was reduced, oversight was reduced or eliminated and not replaced with sufficient "insight," and SE reduced to virtual nonexistence. As a result, billions of dollars of space hardware was lost to failure, as shown in Figure 1.2-5. Both NASA and DoD are currently in the process of reverting to proven practices.

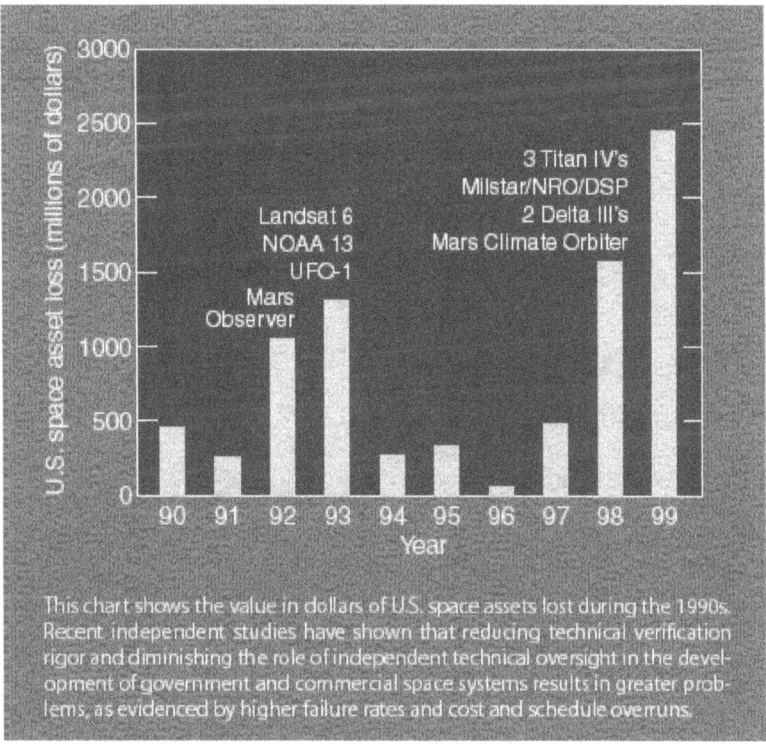

Figure 1.2-5 A Successful Strategy for Satellite Development and Testing
W. H. Tosney & S. Pavlica, Aerospace Corporation Crosslink, Fall 2005

7. Software Development Processes: As reliance on software in space systems increases, so does the incidence of failure caused by its malfunction, as shown in Figure 1.2-6. The role and causes of software failures and anomalies, as well as technical development deficiencies, in spacecraft accidents is well documented in the "Role of Software in Spacecraft Accidents."[7] Examples cited are poor or missing specifications, unnecessary complexity and software functionality, inadequate reviews, ineffective systems safety engineering flaws in test and simulation environments, and inadequate human factors design. The author cites both management and organizational factors, such as diffusion of responsibility and authority and poor information flow as sources of software failures and anomalies. This reference also attributes many of theses factors to a "Faster, Better, Cheaper" acquisition approach.

---

[7] Levenson, Nancy G., Role of Software in Spacecraft Accidents, Journal of Spacecraft and Rockets, Vol. 41, No. 4, July-August 2004

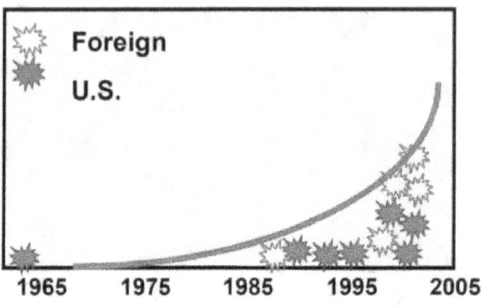

Figure 1.2-6 Flight Failures with Software as Major Cause

8. Following Proper SE Practices: In the last fifteen or so years, Mishap Investigation Boards (MIB) have often identified failure to follow proven and established SE processes as an underlying cause of flight failures. For any new acquisition, it is imperative to recover and use processes that have been demonstrated to achieve mission success, and identify and recruit management and technical staff that have experience designing, producing, and operating space missions using established SE processes. Succeeding sections devote significant explication to the SE process and its application in light of these lessons learned.

9. Failure Causes: Figures 1.2-8 and 1.2-9 show the distribution of failure causes from two different data sets. The distribution of failure causes between the two reports varies because of different missions in the data sets and differences in the classification categories. However, both sets indicate that design, environment, parts, and workmanship/quality make up a majority of the failures causes. Since the available data on parts and workmanship failures are insufficient to differentiate between generic and random causes, it is important to understand the nature of these failures and how they can be prevented or mitigated. Design problems are most likely generic, since they will be present in all units. Similarly, problems related to the environment are generic to all units as they are rooted in the inability of the system as designed to function in the actual space environment. Workmanship problems can be due to either random escapes or problems with the manufacturing processes or paperwork potentially making the cause generic. Similarly, parts problems can be because of a random defect that escaped detection or misapplication (design or part characterization) resulting in stress and early failure. Section 2.2 discusses how the system design might respond in a tolerant and robust fashion to both generic and random failures through the use of redundancy, backups, or other means appropriate to the failure.

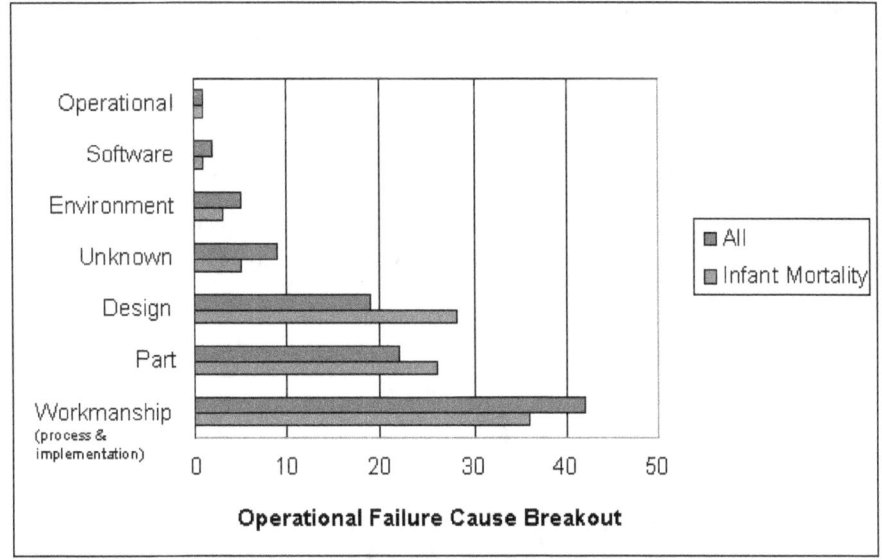

Figure 1.2-7 Orbital Experience from an Integration and Test Perspective
Journal of the IEST, W. F. Tosney & A. H. Quintero, Nov./Dec. 1998

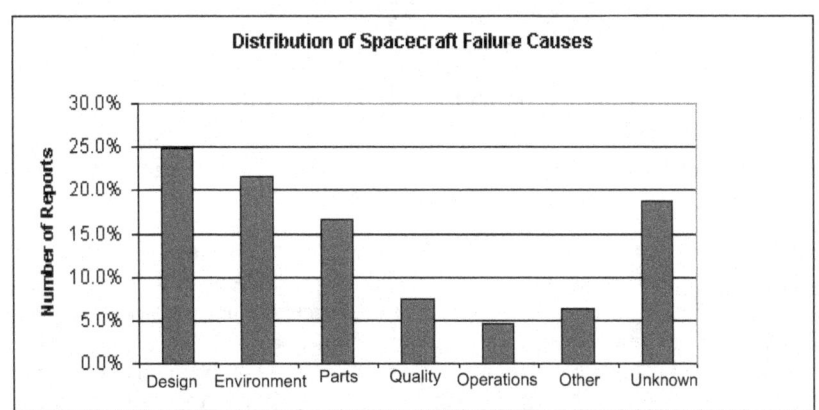

Figure 1.2-8 Reliability Prediction for Spacecraft, RADC-TR-85-229
Rome Air Development Center, H. Hecht & M. Hecht, December 1985

10. Historical Trend of Subsystem Failures: Historically, no subsystem has been immune from failures. One of the questions asked of this report seek to identify if any subsystem has reliability or safety trends that warrant different approaches during design. One caution inherent in the interpretation of historical data is that if subsystem does have a failure, processes are put in place to rectify the problem.

Figure 1.2-9 depicts the distribution of failures to spacecraft subsystems. The study shows that no subsystem has an insignificant number of failures or is immune to failure. Therefore, any subsystem has the potential to result in mission failure. Sub-systems that have safety critical functions need the appropriate attention to meet safety and reliability requirements.

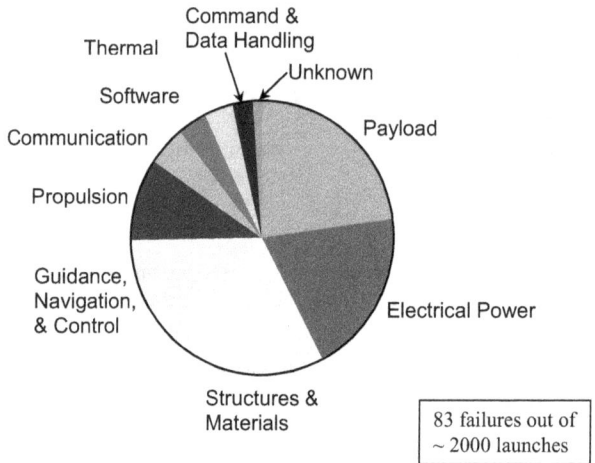

Figure 1.2-9 Analysis from Aerospace Corporation Space Systems Engineering Database (SSED)

11. Reliability and Maturity: Figure 1.2-10 plots an analytical estimate of reliability and probability of success based on flight experience. Reliability improvement is due to accumulated flight experience and the discovery and correction of unexpected failure modes. As the number of flights increase, uncertainties in the operational environment are reduced, how the system interacts with the environment becomes better known, and generic problems are corrected. These factors contribute to reliability improvement. An underlying assumption for failure-free systems is that one failure would occur in three times the number of failure-free flights.

When modeling reliability and considering failure rates, one must take into account the fact that the early flights will be less reliable than a mature vehicle until latent generic issues are corrected. The first flights have a higher failure probability until uncertainties are reduced and "unknown unknowns" become known. Section 2.2 discusses techniques for mitigating the consequences of unexpected failures and Section 3.4 discusses reliability modeling.

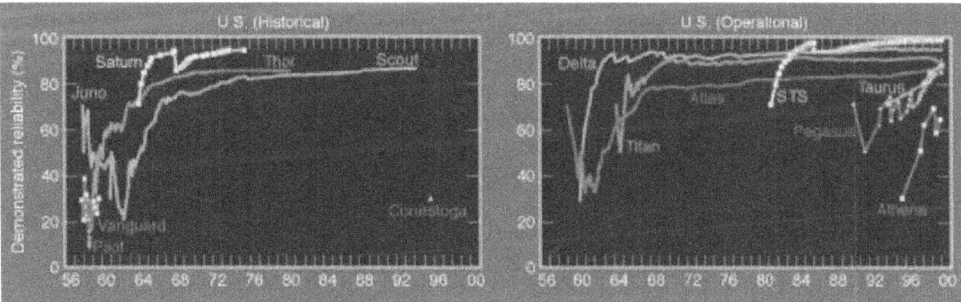

Figure 1.2-10 Demonstrated Launch Vehicle Reliability Improvement with Maturity, Space Launch Vehicle Reliability, I-Shih Chang, Aerospace Crosslink, Winter 2001

History shows that safe and reliable spacecraft can be achieved. Following proper SE practices and understanding the lessons learned over time by the space industry are critical in meeting this goal. With the understanding that failures cannot be totally avoided, it is paramount to put into place an approach that emphasizes safety and reliability in the design. The remaining sections of this report will discuss the key considerations during architecture development and multilayered approach designed to assure the system works as intended.

2.0 System Engineering with a Safety and Reliability Focus

SE plays a critical role in ensuring a system's ultimate safety and reliability. Project teams can achieve these goals by maximizing the leverage and power of the early iterative design cycles conceiving the design, following proven practices for implementing the chosen design, and being alert to warning signs and risks.

Section 2.0 describes a method for achieving these goals from a top down perspective driven by mission objectives and risk. Figure 2.0-1 demonstrates how the topics and key considerations, described in the five subsections of Section 2.0, fit together to achieve safe, reliable, and failure tolerant systems.

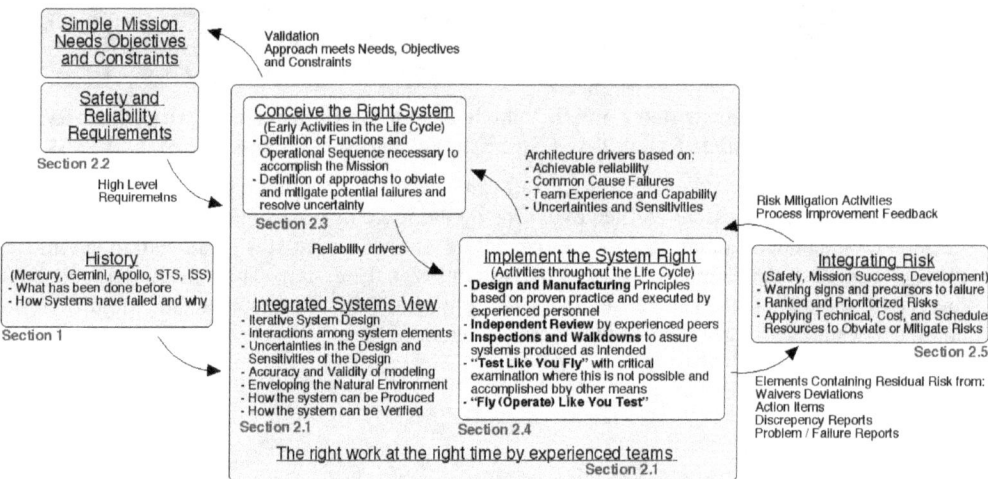

Figure 2.0-1 Conceiving the Right System and Implementing the System Right

Safety and reliability as presented in this report has a broader meaning than just crew safety. It includes mission reliability and "defect free" quality hardware and software.

This report distinguishes between safety and mission success to provide a consequence-based focus that is useful during design, production, operations, and overall risk management. Distinguishing between safety and mission success is important when deciding when the mission should be aborted to preserve crew safety and when efforts to preserve mission success may adversely impact safety.

A safe system ensures the survival, health, and well being of the crew during nominal and off-nominal operational scenarios. It also provides strategies to avoid or deal with unsafe conditions and applies margin to the system to prevent the exceedance of limits that may result in harm to the crew.

A reliable system assures mission success by functioning properly over its intended life. It has a low and acceptable probability of failure, achieved through simplicity, proper design, and the proper application of reliable parts and materials. In addition to long life, a reliable system is robust and fault tolerant, meaning it can tolerate failures and variations in its operating parameters and environments.

A system's reliability is strongly dependant upon its design and is not assured by quality alone. Quality is a characteristic that a system possesses at a fixed point in time, usually at the beginning of life. While quality is necessary for reliability, it is not sufficient.

Reliability includes the following characteristics:

- Free from design or workmanship flaws effecting safety and mission success to the extent practical given available resources
- Tolerant of remaining defects should they surface over time
- Robust to unexpected interactions and environments as they surface over time

Safety and reliability objectives often work together but can also compete. The two work together when margins are added to the system to ensure its continued operation from both safety and mission success perspectives. They compete when safety objectives seek to prevent a hazardous condition that also interrupts mission success. For example, a human rated system would set safety limits, such as engine red lines prior to the point of failure in order to allow a crew abort scenario. An early or false abort may occur at the expense of mission success.

Objective and Overview of the SE Process

The objective of SE as stated in the NASA Systems Engineering Handbook[8] "is to see to it that the system is designed, built, and operated so that it accomplishes its purpose in the most cost-effective way possible, considering performance, cost, schedule, and risk." The NASA Systems Engineering Handbook, SP-6105, provides a solid SE framework and provides an excellent reference for engineers and managers and content details will not be repeated here. This report focuses on the aspects of SE that effect reliability and provides additional details not found in SP-6105 on how reliability goals influence SE activities. Reliability is closely tied to safety and mission success. The goal is to ensure the system functions and safely performs as intended over the life of the mission.

Defining the "Box"

The limits for an iterative system design loop can be thought of as the side walls of a three dimensional box. Credible solutions must exist with cost (schedule), (technical) performance,

---

[8] NASA Systems Engineering Handbook,SP-6105, Shishko, Robert, June 1995, pg. 6

and risk constraints. See Figure 2.0-2 showing the "Box" often utilized to indicate whether a given solution is "inside" or "outside" the "Box."

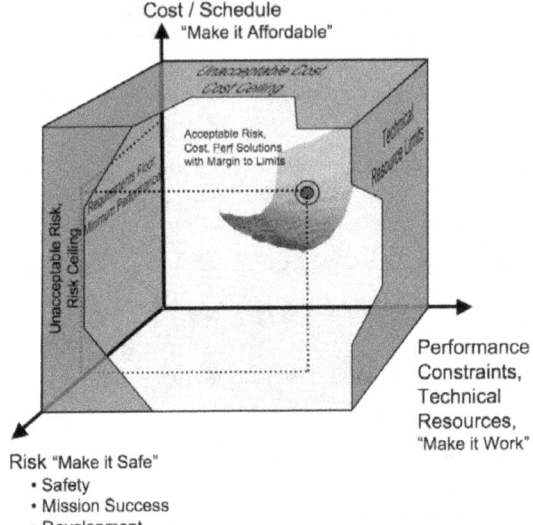

Figure 2.0-2 Trade Space and Constraint "Box" Showing Alternatives as a Surface with Selected Solution

Visualizing solutions along cost (schedule), (technical) performance, and risk axes indicates that each of these parameters affects the other. Notice that this formulation allows for multiple cost effective solutions. Solutions are shown as a surface inside the box constraint walls of figure 2.0-2. The Systems Engineers Dilemma, taken from Section 2.3 of SP-6105, indicates at each cost effective solution to reduce:

- Cost at constant risk, performance must be reduced
- Risk at constant cost performance must be reduced
- Cost at constant performance higher risks must be accepted
- Risk at constant performance higher costs must be accepted

In this context, time in the schedule is often a critical resource; this means, that the schedule behaves like a type of cost.

The desire is to find solutions "in the Box" leaving sufficient margin to the limits to allow for growth. How to get there is the challenge. System engineers must focus analyses and trades on "differences that make a difference" relative to mission objectives, rather than a prescribed level of fidelity, or modeling capability. Key drivers that often "make a difference" are detailed in the following sections. When solutions are "outside of the box," it is quite difficult to eliminate elements of the system to get them "inside the Box." Eliminating system elements threatens the cohesive nature of the system. It is therefore often advantageous to start with a minimal system and add to it only in the deficient areas or in areas where enhancements provide value needed by the system.

The role of the systems engineer is to make sure that solutions work, are safe, reliable and affordable, and that the trade between cost effective solutions are communicated to management. SE must be performed in an integrated fashion to achieve cost effective solutions. Trades on cost, performance and risk must be integrated, or the resulting requirements and designs will likely be sub-optimal or non-realizable. Programs that do not have an integrated picture of the system are doomed to be overtaken by events and become mired in problems with unacceptable solutions.

Achieving a safe and reliable system requires a focused effort at every phase of the life cycle. This entails performing the right work at the right time with the right people, establishing clear mission objectives, a safe human rating requirement, conceiving a balanced design of inherent reliability, like and diverse redundancy, and following sound design and implementation processes with emphasis on safety and reliability drivers at each life cycle phase.

2.1 The Right Work at the Right Time with the Right Teams

SE's task is to find the best balance between program element responsibilities in terms of cost, schedule and probability of functional success. Key participants are designers, crew and ground operators, producers of hardware and software, mission planners, and those responsible for preflight preparation of teams and flight systems. The goal is to design a total system that has the capability to perform initial missions, gracefully accommodate early flight uncertainties and provide high confidence in the ability to safely return crews to Earth from any point in the mission; all within cost and schedule constraints.

2.1.1 Defining Program Needs, Objectives, and Constraints

Defining a clear and simple set of prioritized program needs, objectives, and constraints, including safety, can do much to simplify the system and ensure its safety and reliability. The more simply and clearly the mission objectives are defined, the easier the job of validating the requirements and system and keeping complexity under control.

There are two primary reasons for humans to be in space: to get from one place to another and to apply their unique intellectual capabilities and adaptable dexterity. Humans in space require special consideration because they are relatively fragile, volumetrically inefficient when compared to machines, and require uninterrupted life support, rest and specialized human-machine interfaces. In practice, supporting humans in space represents a significant increase in space system complexity, mass and volume.

Humans in space also significantly alter spacecraft and mission design practices since the safe return of humans is always the primary objective. Therefore, system designs and concepts of operations must place a higher priority on preserving the capability to safely return the crew than to accomplishing the planned mission objective.

The primary objectives of the spacecraft and its related systems are derived from the mission that is to be performed. If the mission description is expansive and ill defined it opens the door for the attachment of all sorts of additional requirements that will eventually produce a system that either cannot be built or, if built, cannot be operated with a high degree of confidence. If the mission objective is kept simple and easy to describe, then the mission objectives can be used to defeat unnecessary requirements and capabilities and the overall simplicity of the resulting system can be defended (in this case simplicity means as simple as possible given the complex nature of the problem).

Apollo is an excellent example. A senior Program Manager, interviewed after the successful conclusion of the Apollo program, related the following commentary, "The mission was easy to define: Man, the Moon, this Decade." So he was always able to able to test the need for each proposed requirement by asking, "Does this help get a man on the Moon in this decade and return him safely home?" If the answer was no, then the design would continue without it.

However, consider the implications of the following mission statement, "Develop a versatile system that is capable of being adapted to perform all plausible, but undefined, future missions." While such a statement may seem facetious, it is very similar to guidance provided at the beginning of STS development. With such guidance, almost any requirement can be defended and the design will grow in complexity without bounds.

Most programs have mission statements that are somewhere in between these two extremes and often require interpretation. As a result, they are subject to mission creep and the Program Office will have a hard time defending decisions that limit capability or reject proposed requirements. When this happens, the results are increased weight, cost, schedule and complexity. Unfortunately, complexity is almost always the antithesis of reliability.

2.1.2 Organizing and Managing the Program

Having established an appropriate objective, the next most important step is personnel selection. To understand the importance of these positions it is necessary to consider the tasks of Program Management, the responsibilities of SE and the relationship between them. These roles are, in fact, two sides of the same coin. Program managers emphasize outcomes and systems engineers the means to achieve those outcomes. Both require vision, not of what to do, but of how to get from here to there with available resources and within real world constraints. In the post Apollo era, economic considerations have become as important as technical elegance and make SE responsible for not only performance and reliability, but also for the minimum cost and schedule. Program Management, then, has responsibility for effective execution. Both communities must develop an intimate familiarity with the mission objective, physics of the task, state-of-the-art of the technology employed, concept of operations, maturity of the design and, most important, the professional maturity of government and contractor teams.

The most effective means of producing successful systems is to put properly prepared and professionally competent managers in charge, provide them with adequate resources, challenge them with a difficult but doable schedule, and give them sufficient authority to be held personally responsible for results. Perhaps the most significant factor in development programs is that these managers must intend, and be expected, to remain responsible for results from day one through commitment to flight. Experience shows that competent people who accept personal responsibility for the final outcome will do more than anything else to find the right balance among the plethora of competing demands that must be addressed during development.

Executive management has three critical responsibilities in making this process work.

- Selecting the right objectives, organizational structure, and key personnel is first.
- After selecting the right people, they must ensure that each manager's authority and responsibility are unambiguous, co-aligned, and supported.
- Finally, having delegated authority and responsibility to a team that will be working under intense pressure for years, management must provide a safety net by maintaining discipline in execution of the Program Plan, ensuring transparency and providing timely independent assessments of progress. Such activities are not bureaucratic, but enable the team to operate aggressively while knowing there will be a second set of eyes to help them avoid errors of commission and omission. Through formal reviews and informal communications the executive leadership will gain insight into the state of the program and the health of the team that is executing it. In so doing, the probability of identifying and correcting the inevitable mistakes will increase, thereby protecting the team, the mission and the agency.

2.1.3 Planning and Pacing of Work throughout the Life Cycle

An important element in managing the Program includes applying appropriate focus to the early phases of the life cycle. Figure 2.1-1 provides a notional example of the power and leverage present in the early phases of the systems life cycle where alternatives and changes are easy to consider. Changes early in the life cycle are easy to implement at low relative cost. As development progresses, changes become harder and more costly to implement. Post-launch changes and fixes become extremely costly.

Figure 2.1-1 shows that decisions made early in the life cycle set the stage and define the destiny for the project going forward. This emphasizes the importance of architecture trades early in the life cycle when there is time and opportunity to change. Reliability analyses performed late in the life cycle cannot effectively drive system architectures and conceptual design. However, reliability analyses performed early in the life cycle can drive the architecture.

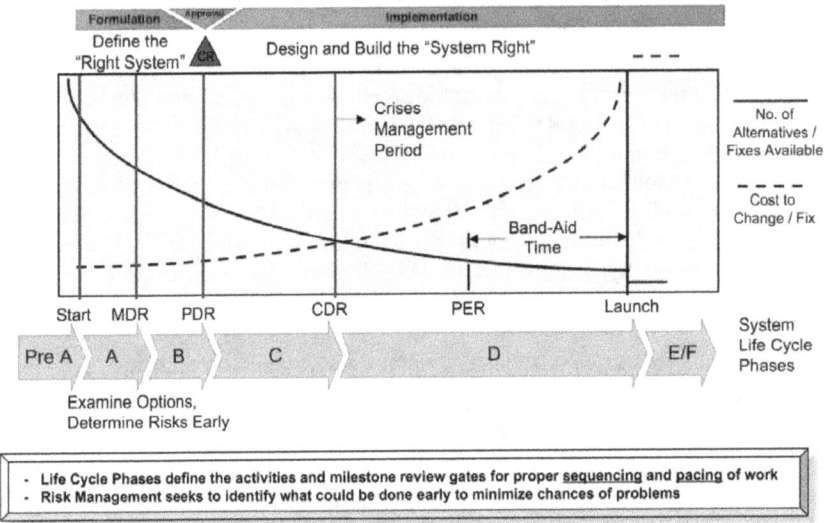

Figure 2.1-1 Planning and Pacing of Work

Another view describing the importance of early design work is shown in Figure 2.1-2 that indicates over 50 percent of the costs for a program is committed by approximately 10 percent of the early design work, and over 80 percent of the total costs are committed after 50 percent of the cost is incurred.

Given the importance of the early work in establishing safety and reliability, Section 2.3 of this report describes a methodology for performing these critical activities.

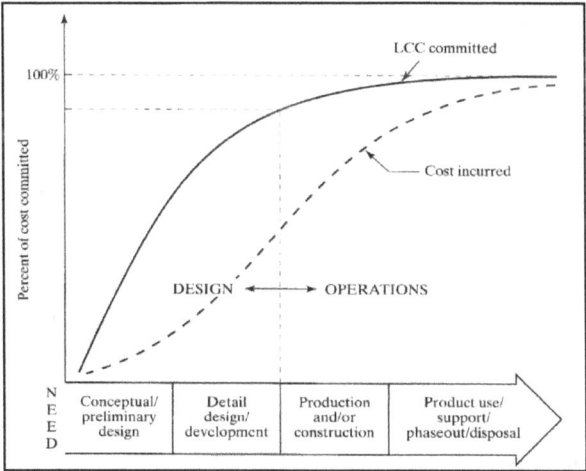

Figure 2.1-2 Cost Committed versus Cost Incurred
Symon & Dangerfield, 1980 and Thuesen & Fabrycky, 2001

2.1.4 Teamwork for Producing Safe and Reliable Systems

The development of an integrated team is especially important to the SE process. Safety, reliability and cost analysts must be included in the systems team along with the designers who are familiar with the systems capability and reliability drivers for their systems. Integration of NASA personnel in these teams is essential to provide real-time feedback of customer safety and reliability considerations. Safety problems identified after the design is complete puts managers in a position of potentially sacrificing safety, adding to mission cost, slipping the schedule, or all of these with resulting negative impact on the program. Designers will respond to the challenge of designing a reliable system if they receive the proper support. While independent oversight of these aspects is needed, these processes cannot be effective if separated for the design team.

It takes a multidisciplinary team to produce and prepare systems that operate in the complex and high-risk space environment. It is necessary to follow an orderly and systematic process to produce safe systems, but following this process is not sufficient to produce a safe system. Project Management, Engineering, and S&MA teams need to work together to design and develop systems that work, are safe, and affordable. These team members back each other up when identifying and correcting potential problems during development and operations.

Safety and Reliability Analysts

The whole project team is responsible for safety and reliability. It is important that design teams have organic safety and reliability experts to help design safety into the system, and develop test programs. The design team itself should perform reliability and safety analysis functions to achieve a safe and reliable design. To assure that schedule and cost issues do not override reliability and safety, the team needs additional support from independent analysts to verify safety and reliability requirements are met. These external members provide peer review and external support for the internal safety and reliability experts. The external safety and reliability requirements verification team should participate in design efforts, test procedure development, and review test results if necessary. Most of the effort put forth to assure reliability and safety should be within the design team itself, with the external team providing an audit and review function.

Team's Safety Focus

The organization should support and encourage a safety focus among team members including suppliers. Teams focusing on safety and mission success[9]:

- Place value in safe and reliable hardware
- Keep curiosity, skepticism, and imagination alive to ask important "What can go wrong" type questions while preventing undesirable effects of complacency and infallibility.
- Know that it takes rigor and discipline to achieve safety and mission success and expend energy to achieve it.
- Stick to tried and proven methods, values, and principles that form the foundation for the processes that achieve safety and mission success.
- Resist being easily swayed by new "fads" or external pressures to change.
- Establish and reinforce teamwork through a "multilayered approach" or a "defense in depth" for mission success.
- Understand that "Risk Management" is not a separate discipline from Engineering, Mission Assurance or Project Management; all team members practice risk management.
- Understand that "Risk Management" is a technique performed by everyone on the team; not something left to "outside analysts" who are called on to "save" the team from making mistakes or make up for lapses in engineering and management.

Multilayered Approach

Space mission complexity along with the uncertainties in system performance argues for a multilayered approach, a defense in depth, to increase the likelihood of fielding a good design and identifying/correcting potential problems before they surface in flight. Team members not only support each other, but also provide multiple opportunities to implement the system right.

---

[9] For additional information: Schmidt, John K. Capt, Characteristics and Assessment of High Reliability Organizations, MSC USN Naval Aerospace Medical Institute, and CAIB staff member

The effectiveness of the safety focus depends on the following factors depicted as the supporting poles in Figure 2.1-3 and described below.

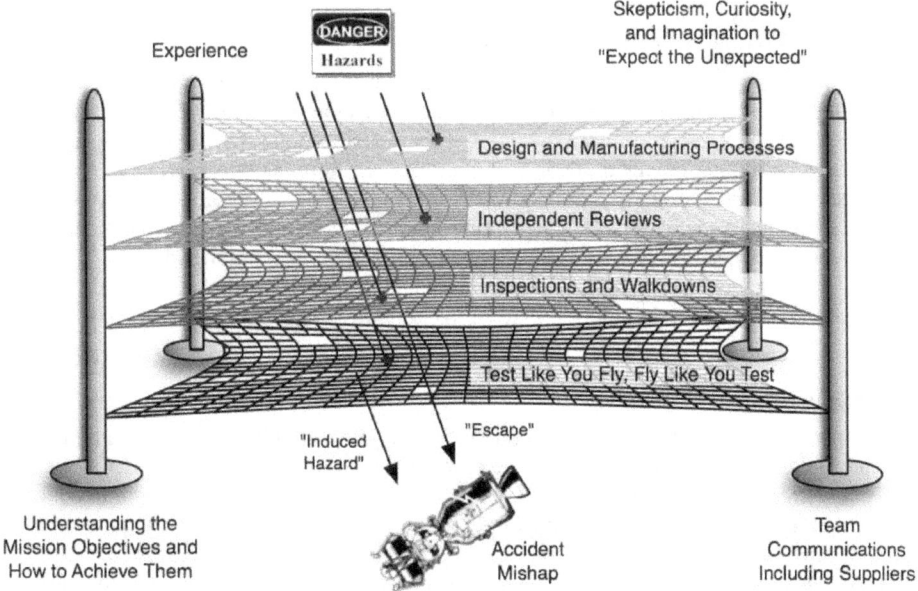

Figure 2.1-3 Multilayered Approach to Produce a Safe and Reliable System

- The team's experience and their access to experienced personnel when needed. Experience allows basic principles, proven processes, and valuable lessons from the past to be applied. Experience is also increased in a learning organization that evolves its methodologies and processes based on feedback from failures, off-nominal conditions and risks. Section 2.5 describes the importance of collecting and acting on off-nominal conditions and risks.

- Understanding and accepting the mission objectives and appreciating what can go wrong along with the consequences.

- A focus on safety and reliability with the curiosity, skepticism, and imagination for what might go wrong. Curiosity: questioning how systems ought to work, finding out why they do not work as expected, following the scientific method. Skepticism: questioning and finding engineering rationale for solutions to assure they are the right ones, keeping complacency under control. Imagination: Astronaut Frank Borman in his 1967 testimony to Congress stated that the cause of the Apollo 1 fire was a "failure of imagination." No one, NASA or the Contractor, thought about the fire scenario in the spacecraft during the "plugs out" test.

- The team should ask, "Why is it safe?" throughout the program life cycle. Affirmative answers to these questions create a preponderance of evidence the system is safe to fly. Concluding an affirmative in the absence of contrary evidence ("It has never failed before.") must be avoided. Long strings of successes can unintentionally lead to misunderstanding or underestimating risks inherent in the system.

- Free and open discussion of safety issues among all team members and suppliers is critical to assuring that all parts of the system receive consistent attention to safety. Communication of concerns and issues should freely flow up the chain to allow a system level view and a considered response. No team member or organization should unknowingly end up inducing a "weak link" into a system's safety or mission success. Extensive communications also are required among the subsystem and discipline teams as they trade and evolve the design. Team members need insight, transparency, and penetration into supplier designs, manufacturing, quality, and test effectiveness and need to develop a "trust but verify" atmosphere focused on the safety and success of the mission even if some of the hardware is COTS and/or has proprietary aspects.

Space Flight Awareness Programs

Space Flight Awareness Programs (SFA) provide an important reminder that all team members are responsible for safety and reliability. The unwritten contract described on the SFA Web Site says, *"No matter how well spacecraft are made, safety margins in space travel will always be small. A space vehicle is only as reliable and safe to fly as the human care that goes into its creation. For that reason, each individual associated with human space flight is party to the unwritten contract, "Flight Safety and Mission Success."* SFA uses a variety of motivational awards and incentives, along with active education, to effectively remind the entire Government/Industry Team of their role in achieving flight safety and mission success.

Human space projects have utilized a form of the SFA Program as a way to mitigate risk and assure safety and mission success. Aspects of SFA, or its predecessor, Pilot Safety, have included traditional quality control techniques. Examples include parts screening or special handling of critical parts.

The future of space flight brings new opportunities and challenges for the SFA Program. To continue to be effective, the Program must keep pace with an ever-changing environment of people, systems, and processes that design, build, fly, and support human space flight yet remain grounded in the understanding that safety and reliability are the responsibility of each individual.

2.2 Defining the Requirement for a Safe Human Rated System

Humans in space require special consideration because they are relatively fragile, volumetrically inefficient when compared to machines and require uninterrupted life support, rest and specialized human-machine interfaces. In practice, supporting humans in space represents a significant increase in space system complexity, mass and volume.

Humans in space also significantly alter spacecraft and mission design practices since the safe return of humans is always the primary objective. Therefore, system designs and concepts of operations must place a higher priority on preserving the capability to safely return the crew than to accomplishing the planned mission objective.

A safe and reliable human rated spacecraft system provides a reasonable assurance that the crew survives nominal and off-nominal operational scenarios. NASA-STD-3000 Man-Systems Integration Standards provides a set of requirements that assure crew survival in the space environment including a breathable atmosphere, safe thermal and g-loads, safe radiation environment, adequate nourishment, and personal health and hygiene opportunities.

Assurance of crew survival during off-nominal and failure scenarios leads to requirements that seek to prevent faults and assure crew safety in spite of faults. History, as summarized in Table 1.0-1, shows a combination of three basic methods for specifying safety:

- Specify fault tolerance (i.e., X fault tolerance to loss of crew, Y fault tolerance to loss of mission)
- Bound the probability of failure and assess the vulnerability of the design to existing internal and external hazards
- Require adherence to practices that are known to produce safe systems

Choosing a single method from the three above cannot, by itself, ensure a safe system. Potential pitfalls of taking each of these kinds of requirements individually are:

- Specifying fault tolerance alone can force the addition of hardware that can decrease reliability due to added complexity.
- Limiting the probability of failure alone can result in unrealistic probabilities and a potentially false confidence that the system is safe. Modeling is often imprecise and can contain a high degree of uncertainty.
- Specifying adherence to practices or recipes alone can miss an important aspect not totally codified in the process.

The challenge is to specify the requirement in a manner that achieves the intended result, i.e., a "safe" system. No single technique of providing redundancy, or backup systems, generating a fault tree, FMEA, or hazard report, tracking critical items lists, or calculating the probability of loss of crew assures actual crew safety. No single activity such as design, independent review, inspection, or testing can by itself assure that the system is safe.

Ultimately, the requirement for a "safe" system must include and balance requirements in all three of the following areas (see Figure 2.2-1):

1. Fault tolerance at the Program level assessed from a top down functional perspective and flow it to lower levels only after it is optimized from the total system perspective. Where meeting the desired fault tolerance by design is not practical the acceptance rationale includes a design

for minimum risk supported by a quantitative risk analysis described below to assess residual risk.

2. The objective of bounding the probability of failure is to encourage a thorough investigation into risks including uncertainty and common cause such that the system design decisions and underlying risk analysis can be defended. Estimating the likelihood of failure challenges the design team to achieve a deeper understanding of the system and its environment, and provides a way to make trades when protecting against worst case assumptions cause a design to become overly complex or over weight. The value of a probability estimate is not so much contained in the absolute number but in the thorough investigation, debate, and discussions by designers and operators about controlling the potential for failures based on their likelihood, the history of similar systems, and the uncertainties inherent in the system design. The analysis of the system design must consider the integrated whole and include a top down assessment. The analyses are most useful for evaluating and comparing design and operations alternatives and validating the chosen system design. The purpose of the Probability of failure (Pf) requirement is to encourage a balanced design maximizing safety at the system level. The quantitative analysis helps designers identify weak links and safety drivers and provides rationale where a design for minimum risk approach is taken. To ensure valid comparison correct statistical methods should be used to determine probability of failure and include all available data sources.

3. Identify and follow proven practices and processes for design, manufacturing, independent review, inspection, test, and operations. Processes can achieve consistency and provide a method for following through on details necessary for safety and reliability. Disciplined adherence to critical processes is important for achieving the safety and reliability the designers intended.

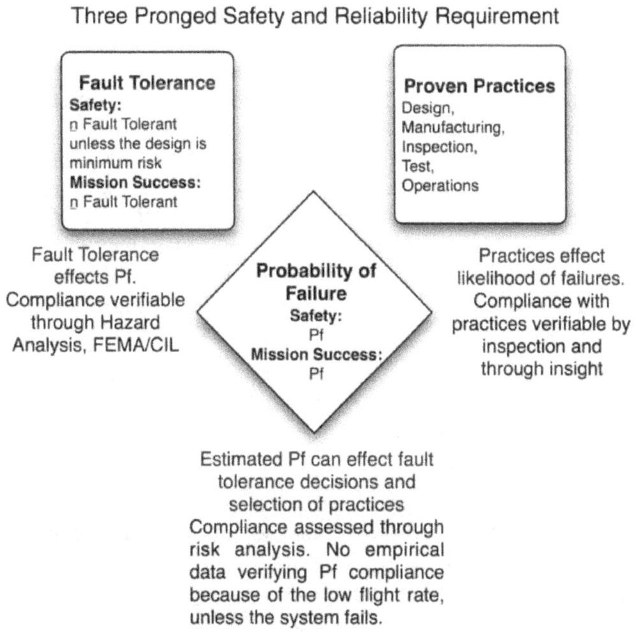

Figure 2.2-1 Three Pronged Safety and Reliability Requirement

Taken together these three elements of the reliability requirement synergistically drive the development of a safe system. A system designed with the required fault tolerance, supported by probabilistic reliability and risk analyses, produced and operated with attention to the appropriate details ensures a safe system.

Risk Assessment considers the design, its fault tolerance, Pf, uncertainty, common cause effects, the operations concept, and all the other sources of risk to the mission including the environment, hardware, software, operations, crew, and the ground. The results of the risk assessment are evaluated not only by the development team but also by independent peer reviewers.

Assessing safety and reliability starts at the onset of the program with high level functions and continues throughout the life cycle as the design matures and system elements are produced, integrated, and prepared for flight.

2.2.1 Fault Tolerant Requirement

The two-fault tolerance requirement along with a design for minimum risk option (NPR 8705.2A Human Rating Requirements) for human-rated systems is established to ensure that systems are safe. This is a good philosophy given that: 1) each of the two failures is unlikely on its own over the mission life, 2) taken in combination two failures are extremely unlikely, and 3) the system is not susceptible to a generic failure or other failure that defeats the system's failure tolerance due to a common cause.

Two-fault tolerance is one way to reduce the impact of a failure within a system that can compromise the initial unit and not the back-ups. For two-fault tolerance to be successful, the back-up systems must not be compromised by the original failure mode, cause, or hazard. Achieving fault tolerance through redundancy does not provide complete protection since most space system failures are not caused by a single "random" event that causes failure of only a single component. If failure occurs, it is usually due to a process failure, or "unknown unknowns," that will affect multiple components (e.g., manufacturing flaws, design errors, and harsh environments). Refer to Section 1.1 Failure History for information supporting this conclusion.

In some cases, an implementation of a two-fault tolerant requirement can be construed to mean that two-fault tolerance is all that is necessary for the definition of a safe system. Limiting the total system requirement to only failure tolerance can allow a developer to assert that there is a safe system just because it meets a failure counting or redundancy assessment.

It is difficult to "pass-down" a two-fault tolerant requirement to lower levels of the design since a lot of mitigations occur at the highest level where the systems interact. The integrated mission level must establish the overall mitigation strategy and fault tolerance across all available systems, and pass specific fault tolerance requirements (0, 1, 2) down once they have been optimized by the systems engineer.

It is important to recognize that fault tolerance requirement is applied at the functional level, not the hardware level. Therefore, fault tolerance must be considered from the top down in a design. Fault tolerance must be evaluated at the system level with a deep knowledge to ensure that the system will perform in nominal and off-nominal conditions. If two-failure tolerance is applied at too low of a level, within individual elements or subsystems, the resulting addition of mass and complexity into the system can result in a decrease of total system reliability. Evaluation of FT should encourage the use or alternate application of existing functional capabilities before adding complexity to the system in the form of like or diverse systems. For example, the Reaction Control System can provide an alternate method for providing propulsion to a main propulsion system in some instances, and can provide an alternate method of controlling the thrust vector of the engine in addition to engine gimbals.

SE plays a key role in identifying the best strategies for incorporating failure tolerance into the system. SE should apply a conscious strategy for how fault tolerance is attacking flaws in the system, protecting against potential common cause, and "unknown unknowns" (refer to Sections 2.3, 2.4, and 3.0).

It is important to recognize the difference between the goals of mission success and safety. Crewed systems require the crew to return home safely even if the mission cannot be continued. The minimal functions required to return home safely are typically much less complex than those required to perform the mission. This provides an opportunity to use alternative capability present in the design or develop simple systems as an independent back-up. If simplicity is maintained, these systems will be easier to develop, test and verify, and have higher reliability.

Crewed mission systems tend to have a large number of components required for mission success. The additive impact of the likelihood of individual failures is such that single fault tolerance is required to assure that the mission can go on in the face of a single failure. This might require additional systems, redundant channels, and automation for fault detection and isolation. In cases where the likelihood of failure becomes high (e.g., long duration missions) additional channels may be required. These channels can be implemented through on-line or off-line spare, or repair.

The first consideration when choosing either like-redundancy or backup is mission continuance. To mitigate the consequences of the first failure, redundancy should provide full performance to ensure mission success. The system can utilize one, two, three, or as many redundant strings as necessary to reach the operational reliability required as long as these additions are within the performance constraints, provide mission success value and do not compromise safety.

After meeting mission success reliability goals, the consideration of backup legs should be focused towards ensuring safety. The focus shifts to assured safe operability and not on performance. This almost always implies a diverse approach to minimize the possibility of a common cause failure. Examples of diverse systems to ensure safety are an ejection seat or launch escape system versus another engine, a star tracker, and earth limb optical system rather than another platform. The focus also indicates that when there is only one copy of a system element then the emphasis should be on safety and not on operational reliability. Lower performance with higher reliability is to be preferred for safety provided it does not violate the first rule of design, which is "make it work."

2.2.2 Quantitative Requirement and Supporting Analysis

Bounding the probability of failure encourages a thorough investigation into risks including uncertainty and common cause providing justification and rationale so that design decisions can be defended. Estimating the likelihood of failure challenges the design team to achieve a deeper understanding of the system and its environment. The value of a probability estimate is not contained in the absolute number but in the process of developing it. It is the debate and discussions of risks and their likelihood that encourages a thorough investigation by designers and operators about controlling the potential for failures based on their likelihood, the history of similar systems, and the uncertainties inherent in the system design.

The analysis of the system design must consider the integrated whole and include a top-down assessment. The analyses are most useful for evaluating, comparing design and operations alternatives, and validating the chosen system design. The purpose of the Pf requirement is to encourage a balanced design maximizing safety at the system level. The quantitative analysis helps designers identify weak links and safety drivers, and provides rationale where a design for minimum risk approach is taken.

A quantitative requirement bounding the probability of failure can support the fault tolerance requirement. The quantitative requirement along with its supporting analysis allows an assessment of risk and a mechanism to rate and rank alternatives. Quantitative requirements and their analyses need to include all elements of the total system, even those launched on multiple vehicles.

Mission life can drive reliability, design, and test approaches. Planned six-month lunar missions are an order of magnitude longer than Apollo and Shuttle Missions, and the Mars missions will be significantly longer than Lunar Missions. Assessing safety and reliability for these longer missions requires quantitative assessments for safety and mission success as fault tolerance alone may not adequately identify weak points in the system.

The quantitative requirement can take the form of a success or loss probability expressed on a per mission basis. Earlier manned Gemini and Apollo missions used a success probability on a per mission basis. Table 1.0-1 in Section 1.1 shows the different philosophies used to define the reliability requirements.

Military and civil aircraft have used the Probability of Loss of Aircraft (PLOA) expressed per flight hour. Because the risk in manned space missions is extremely dependent on mission event (i.e., launch versus cruise or loiter), hourly risk estimates may be misleading. However, long duration lunar missions and missions to Mars need to consider failure rate per hour in order to estimate a mission level reliability.

The metric of PLOA is usually expressed in terms of failure probability per flight hour; the data in Table 2.2-1 are summarized from Table 1.1-1. What these data indicate is that US-manned space missions have a probability of loss of vehicle per flight hour that is of the same order of magnitude as fighter aircraft missions. Commercial aircraft then have a probability of loss of vehicle per flight hour is two to three orders of magnitude safer than fighter aircraft.

Table 2.2-1 Probability of Loss of Aircraft Summary Data

Vehicle	PLOA Probability of Loss of Aircraft	Flight History
Space Shuttle	7.8e-5 losses per flight hour	2 losses in 25440 flight hours, 1060 days through loss of Columbia
Apollo	Estimated 7.4e-5 to 3.7e-5 losses per hour (assuming 1 failure over 2 to 4 times the number of flight hours)	No losses of vehicles or crews in 6760 system manned flight hours (discounting Apollo 1, which never flew and one loss of mission, Apollo 13, with successful recovery of the flight crew). For the 15 manned Apollo flights (including 3 long-duration Skylab flights)
F16	3.9e-5 losses per hour	from Table 1.1-1
Commercial Aircraft	9.1e-7 to 5.6e-8 per flight hour during 1983-2002	from Table 1.1-1

Establishing Failure Rates and Uncertainty

One key driver in estimating the probability of failure and reliability is the uncertainty in the failure rates. Failure rates are dependent upon the maturity of the vehicle, and thus go beyond the piece part failure rates provided by MIL-STD-217 and other techniques. Flight experience increases knowledge about system performance and allows the discovery of unknowns. Flight performance trends and flight failure history should feed back into design and development processes and safety and reliability analyses. It is important to recognize that estimates and requirements based on mature systems are likely to be optimistic when applied to manned spaceflight with a limited number of missions. In this case, estimates must be made in the context of the overall traffic model that affects system maturity. See Section 3.4.3 for a discussion on modeling maturity.

Tracking and trending of failure rates should be conducted using analytical tools such as Crow-AMSAA[10] and statistical process control (SPC).

Failures can occur due to common cause failures, operational sequences, maintenance and other induced causes not related to piece part failures. Typical reliability analyses do not consider these issues. The systematic approach of a probabilistic risk assessment can consider these if the analysis and design teams work together to explore failure causes beyond piece part failures.

Maturing Estimates along with Design and Operational Maturity

Probabilistic Risk Assessments (PRA), performed along the systems life cycle, provide the quantitative analysis showing compliance with the requirement. To maximize the PRA's effectiveness, it is tailored to the appropriate level of detail given the maturity of the design.

---

[10] Crow-AMSAA model developed by Dr. Larry Crow for the U.S. Army Materiel Systems Analysis Activity which has been incorporated into DoD military handbooks.

Considering all program elements is important when calculating probability of mission success and safety. For example, the ISS row of Table 1.0-1 indicates ISS reliability is dependent upon maintenance/repair through the supply of parts by other vehicles such as Shuttle. Shuttle availability and its reliability then factor into ISS reliability through establishment of maintenance and repair supply lines.

Reliability analysis early in the life cycle includes an FMEA along with fault tree analysis, reliability block diagrams and predictions to help discriminate among design alternatives. Quantitative analysis early in the life cycle can use ESDs and event trees to summarize and quantify the differences. Quantitative probabilistic and uncertainty analyses in addition to associated internal and external phenomenological analyses are used to highlight the risk discriminating differences among design alternatives in design trade studies. FMEAs, fault trees, and reliability block diagrams should be used to support these analyses, but not to replace them.

Traditional FMEAs are component oriented and bottom up so they are of limited use in the early developmental stages. However, a functional FMEA is an appropriate tool that can be used to evaluate potential failure modes even in early development designs. Properly applied, they focus designers and others on avoiding design pitfalls from the beginning of the process.

As the design becomes more defined after the Preliminary Design Review (PDR), then comprehensive FMEAs and quantitative risk models such as detailed PRA are used to better represent the forecasted performance of the selected design alternatives and form the basis for focusing the developmental test program and for tracking the developmental risk.

As development, test, and operational issues surface the quantitative analyses are updated to identify areas of the system potentially threatening the system reliability requirement.

Section 3.0 describes the various reliability analysis techniques and how they are applied along the systems life cycle.

2.3 Conceiving the Right System, Critical Activities Early in the Life Cycle

Early SE activities are critical to defining how the system accomplishes its purpose and how the system responds to the unexpected. Early system design activities have a high degree of leverage as discussed in Section 2.1. These early trades investigate how the system works, how it remains safe, and how it remains affordable. These early trades define how the selected system solution is placed in the Cost, Risk, and Performance "Box" discussed earlier in Section 2.0.

SE activities during Pre-Phase A and Phase A have a dramatic influence on its ability to survive the effects of unexpected differences between the designer's intent and systems behavior, as well as system failures. The system architecture defines how the system responds to unexpected environments, unplanned operational sequences, and failures. The need for a system to cope with "unknown unknowns" is a fact for space systems.

Architectural and conceptual designs define how system/subsystem elements are interconnected and the types of like or diverse functional redundancy available to react to the unexpected conditions and to failures. Allocating functions to subsystems including software and selecting redundancy approaches is best accomplished at the system level in an integrated "systems level" view. It is only at the systems level with subsystem participation that the interactions can be evaluated.

The overall system architecture design activity involves balancing what is wanted and what is not. Assumptions are made early in the life cycle that tend to be linear, based on simple models and simple interfaces. Sometimes initial simplifying assumptions and the real world performance of the system result in unwanted interactions when the product is actually assembled and operated. As the design matures and discipline experts proceed with their detailed design, the physical non-linear realities start to surface, and may add complexity and unintentional interactions. Systems engineers need to revalidate the system design, the compatibility of interfaces and total integrated system performance, safety, and reliability as the design matures.

In addition, the overall design activity involves a margin assessment not only in the traditional areas of mass and power, but also in the robustness of the design against potential threats. The available margin to counter these threats is often the key to crew safety. Margin guards against negative consequences resulting from the occurrence of "unknown unknowns," and these are often the primary threats to the crew because they cannot be specifically anticipated and accounted for in the design.

The primary SE objective is to design the system for performance and mission success through an iterative loop, evaluating risk drivers, and considering alternative derived requirements, design solutions, or operations concepts.

2.3.1 Managing Complexity

System complexity must be minimized since it is the most significant feature of systems that fail. Complexity impedes the designer's understanding of how various system elements might interact and can prevent a full understanding of the integrated system. Human spaceflight operates on the boundaries of technological abilities. It is a highly integrated complex activity that requires the sequential success of a large number of active subsystems all of which are operating close to their limits. As such, a small increase in complexity may have a negative impact on safety and reliability. Complexity is the antithesis of reliability, and should be limited to what is needed to accomplish the mission objective.

Increasing Complexity to Achieve Safety and Reliability

System complexity has a major effect on the system's reliability. Care needs to be exercised when the system complexity is increased in an attempt to improve safety and reliability. System designers need to consider the ultimate effects of complexity on system reliability when additional units or redundancy are added to the system. Predictions are useful for evaluating the relative effects of alternate architectures on system reliability. Redundancy is often mistakenly limited to considering identical unit replication or adding another string or strings. This simplistic approach will often not suffice especially in complicated interacting systems that are weight and cost constrained. An integrated approach considering common cause is needed and described in Section 2.3.2.

Managing and Integrating Pieces into a Cohesive Whole

A common method for managing large and complex systems is to divide the whole into smaller, simpler "manageable" pieces, and allow separate groups to individually produce those pieces. The splitting of the system into pieces must occur from the top-down considering the critical functions necessary for safety and reliability. Engineering managers must have a firm grasp of the risk drivers for their system, even if they are at very low levels of the WBS. The engineering and management challenge then becomes the process of reintegrating the pieces into a cohesive system while avoiding adverse couplings and interactions that may affect safety and reliability.

Simplified designs, models, and interface assumptions made early in the life cycle often turn out to be more complex when actual systems are produced and tested. The role of the systems engineer in integration requires the mindset of a "generalist" who can identify critical functional, physical interfaces, and interactions among tightly coupled system elements. Functional and physical interfaces must be kept simple so newly joined elements of the system that may adversely interact and compromise safety and reliability can be identified.

Design teams responsible for individual system elements must be aware of their system's sensitivities and unwanted interactions with other system elements to understand potential adverse coupling with other systems. It is important for the SE Team to recognize the importance of interaction among discipline engineers after requirements have been allocated and to capture cross-interface information in ICD.

Controlling Implications of New Technology

The mix of new and existing technologies in a design can add to complexity that affects safety and reliability. New technology can improve safety and reliability when carefully selected and applied though new technologies often bring with them "unknown unknowns" that may represent safety and reliability risks.

In cases where new technologies are necessary, the systems engineer must help the design teams assess the maturity and identify potential interactions, along with additional constraints and uncertainty the new technology might introduce. Introducing new technologies may make the system more reliable at maturity, but failures during the maturation process may make the system less reliable when considered over the life of the program. The systems engineer must understand how new technology introduces unknowns into the program, and what can be done to combat them, for example incorporating additional margin, extra testing, alternative flight manifests and concepts of operations.

Applying Heritage and COTS System Elements

Using "Heritage" and COTS system elements are often utilized as a way to reduce risk. However, COTS elements can introduce complexity and risk if they are not applied properly.

COTS products bring with them design constraints, predefined interfaces, and operational constraints that the receiving system must accommodate. COTS elements, especially those with a proven flight track record, can improve safety and reliability, but it is their proper application and accommodation in a new and different application that represents a challenge to the systems designer. For COTS and heritage components, the design focus shifts from having to define the component's detailed requirements (as in newly developed items) to accommodating its constraints.

Ultimately, it is the responsibility of the engineering team to ensure that the benefits are realized and that unknowns are discovered before flight. The promise and advertised benefit of a new technology or COTS elements is often not realized in practice and should therefore be addressed from a risk perspective.

2.3.2 Iterative System Design and Defining the Right Requirements

The iterative system design loop seeks to converge on a set of self-consistent requirements, conceptual design, and baseline operations concept. All three of these together must then meet the project's needs, objectives, and constraints. In fact, users need to participate in the overall SE effort to ensure unique needs are addressed in the overall system design. Including the crew and Mission Operations Division in the early architecture design work is crucial, as they are the actual "users" of the system. Figure 2.3-1 shows the early systems design flow described below.

Iterative system design starts with the team's understanding of the high-level project needs, objectives, and constraints as shown in the box on the left in Figure 2.3-1. These high-level needs, objectives, and constraints form the validation basis for the subsequent system design. Defining a simple set of mission objectives is described in Section 2.1.1. Constraints include technical, cost, schedule, and risk limits used to define the walls of a constraint box shown in Figure 2.0-2.

High-level mission objectives are typically captured as "shall" requirement statements and often captured in the highest Level 1 requirements. Mission objectives including destination, purpose, stay time, crew size, and support requirements often are a "shall" statement defining elements of the constraint "Box." As described in Section 2.1.1, these mission objectives need to be as simple as possible as they form the basis for validation of the design work and ultimately drive the complexity of the system.

It is critical that lower level derived requirements not be written too early. "Shall" statements derived from the top-level requirements (mission needs, objectives, and constraints) should not be written before there is time and opportunity to iterate the design and operations concept along with the requirements.[11] Once "shall" statements are written, it could appear that viable options inconsistent with the requirement statement are precluded. Shall statements written too early may

---

[11] Section 2.5 of the NASA Systems Engineering Handbook SP6105 stresses that *"system requirements, and constraints should be left at a high level for as long as possible so that alternatives can be considered and uncertainties resolved before the most cost effective solution can be obtained."*

unintentionally preclude other viable and safer solutions allowed by an alternate set of requirements. The "right requirements" define the approaches that are consistent with the design and operations concept and have been validated against the mission needs, objectives, and constraints (especially cost). These "right requirements" are best written after a few passes through the iterative system design loop.

After establishing the mission objectives and constraints, it is important that the system designer identify the simplest set of functions necessary to meet the user's needs shown in the next box in Figure 2.3-1. Once the basic functions are identified, operational considerations and needs for each function are evaluated for each mission operational phase providing a detailed operational context for the subsequent design efforts. Each function's criticality with respect to safety and mission success is identified. This distinction is critical for assessing and accepting a function's criticality, appropriate fault tolerance, and probability of failure.

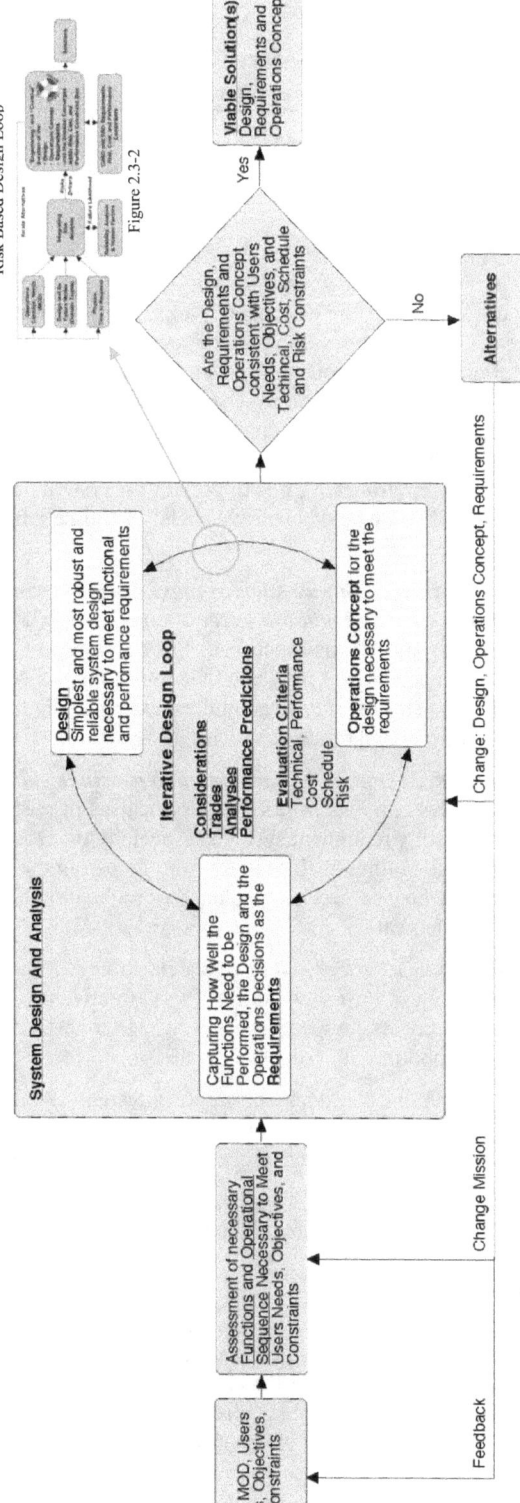

Figure 2.3-1 Early Systems Design Flow

A key requirements driver is the understanding and estimation of the natural and induced environments based on the design and the operations concept. The system designer must develop an understanding of the internal and external environment anticipated during system operation. In addition, the designer must develop protective or mitigating features, to defend the system against normally anticipated excursions in these environments and abnormal excursions in proportion to their risk. The uncertainty in the environment seen by each element of the system drives the margin required to envelop uncertainty. A significant amount of uncertainty can exist in the environment and how it is modeled. To understand how system elements may react and operate in the estimated environment, it is important for system designers to perform sensitivity studies to understand if and where potential "cliffs" or ultimate limits exist in the system.

The SE team must allocate technical resources such as mass, volume, power, fuel, etc., as necessary, to meet the needs of safety and mission success functions. Allocation of scarce resources must consider the importance of the function they are supporting. When resources are constrained, they should be assigned first to safety and then to mission success functions. Therefore, it is critical that the systems team have an understanding of the functions and their criticality before allocating resources and proceeding with design.

Physical implementation of the system must be considered to preclude spatial and physical interactions between elements that can lead to failure. Co-location of redundant units can lead to common cause failures. Separating redundant systems, especially diverse systems, may be required to prevent common cause failures such as leaks from defeating the intended redundancy. Additional iterations may be necessary to address interaction of system elements at its interfaces with other system elements.

Up-front design must consider how a system will be tested in order to accommodate functions necessary for verification. Design can influence the costs and effectiveness of testing and inspection that occur later in the process. Aspects such as "design to test," and "design to inspect/maintain" concepts are important to initial system design.

Many decisions are necessary to define a system's conceptual design. Useful trade techniques and evaluation criteria are described in the NASA Systems Engineering Handbook SP6105 and other references, which will not be repeated here. This handbook is a valuable reference for the systems engineer.

The following section describes an iterative design loop driven by risk assessments.

2.3.3 Risk Based System Design Loop

This report describes a "risk-based" methodology, shown in Figure 2.3-2, to drive a safe and reliable design. Since the early conceptual design has a high degree of influence on the system's safety and reliability, a top-down perspective for the entire vehicle and its mission is necessary. The system design, its complexity, and the nature of the interfaces among system elements as well as the intrinsic reliability of its component elements define system safety and reliability.

Risk-based design provides a methodology to consider risk early in the design when obviating and mitigating risk is relatively easy.

Conceptual designs are driven by the functions necessary to meet mission objectives, the way the functions can fail, and how these failures can be obviated and mitigated. Henry Petroski writes in his book titled "Things Small and Large, *"Things that succeed teach us little beyond the fact that they have been successful; things that fail provide proof that the limits of design have been exceeded. Emulating success risks failure; studying failure increases our chances of success. The simple principle that is seldom explicitly stated is that the most successful designs are based on the best and most complete assumptions about failure."* The risk-based design loop therefore addresses how a design might fail and how resources can be used to obviate and mitigate risk.

An integrated team representing the crew, mission operations, mission design, SE, subsystem domain teams, reliability, human factors, test, safety, and quality assurance are necessary to explore alternatives and select an optimal approach. It is through the participation of experienced team members that diverse ideas and potential solutions are considered and evaluated from a safety and reliability perspective.

The starting point for the iterative design loop is the simplest possible configuration based on the functional block diagram. The team then utilizes the iterative loop to refine the system as necessary to adjust fault tolerance and redundancy as driven by risk.

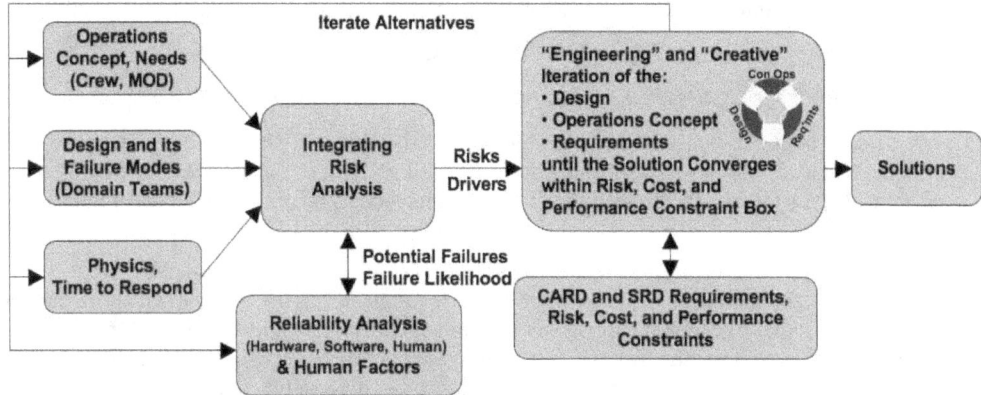

Figure 2.3-2 Iterative Risk Based System Design Loop

Figure 2.3-2 shows the risk-based iterative loop assessing risk through the use of reliability analysis and is described below. The iterative loop seeks to "make it work," "make it safe and reliable," and "make it affordable."

The team combines the operations concept, along with the characteristics of the design, considering both normal function and potential malfunctions and failures. These are identified by a functional FMEA, fault tree analysis, and the physics of the situation including hazards and the time to respond to failures as inputs to an Integrating Risk Analysis as shown in the left hand boxes in Figure 2.3-2.

Next, an Integrating Risk Analysis is used to identify the risk drivers shown as the middle box in Figure 2.3-2. An ESD is used as an example in this report for the integrating element. ESDs can be used to integrate driving characteristics of the operations concept, design, and its failure modes, physics of the situation, and reliability analysis to identify the risk drivers for the solution under consideration. The ESD provides a method for both a time ordering ("sequence") along the horizontal axis and a functional relationship along the vertical axis. In this manner, the ESD serves as a powerful technique to assess successful and failure events.

Other integrating elements or diagrams can be used, such as functional flow diagrams or fault trees for each major functional or operational configuration. However, careful consideration must be given to include and capture the operational sequence and operations concept with these techniques.

The Integrating Risk analysis:

- Considers nominal Mission Sequence of Events or Operations Concept for accomplishing the mission and allows evaluation of failure scenarios.
- Identifies necessary functions and a conceptual design
- Identifies failure modes of functions and external hazards
- Estimates likelihood of failures and their consequences
- Identifies system passive/active response to likely failure modes
- Based on the response of the system, identifies risk drivers. Make note of how much risk is "set aside" because it falls into "unlikely, assumptions, or sensitivities;" if what gets set aside gets too big, then revisit the drivers and include some of the 'unlikely' in this analysis.

The next step shown in the right hand boxes in Figure 2.3-2 takes the resulting risk drivers and iterates the design, operations concept, and derived requirements as necessary to either obviate or mitigate the risks. The optimum approach is to eliminate the risk by altering the design or operations concept if possible. Approaches to mitigate risks included using existing system elements in alternate ways, addition of redundancy or other diverse functional backups, as well as designing for a highly reliable design for minimum risk.

Each iteration cycle considers an alternative. Alternatives should be plotted as a surface within the constraint box shown in Figure 2.0-2. The ultimate solution is then chosen from the family of alternatives considered. Alternatives considered, but not chosen, are also important in shaping and defining the ultimate solution and providing the justification and rational for the solution.

As an example, Figure 2.3-4 shows the progression of alternatives considered during the Exploration Systems Architecture Study (ESAS) launcher study. Figure 2.3-5 shows how uncertainty estimates are useful in determining the significance of the differences between alternatives.

Below is one path applicable to the upfront work in the life cycle that can lead to a safe, reliable and affordable design.

Step 1 Define needs, objectives, and constraints in clear and simple terms, and then capture them as the high-level requirements. These form the validation basis for the subsequent work. The constraints can be visualized as the boundaries of a box as shown in Figure 2.0-2. Subsequent steps seek to find solutions within box constraint boundaries and select one for production. Alternatives should be plotted against constraints forming a "surface" from which a solution is selected.

Step 2 Define the minimum set of functions necessary to accomplish the mission objectives.

- Identify and describe the functions the system must perform from a systematic top-down perspective in order to fulfill mission needs and objectives.

- Clearly identify and distinguish functions necessary for safety and mission success. This distinction is critical for assessing and accepting a function's criticality, appropriate fault tolerance, and probability of failure.

- The identified critical functions should be used to set up the PBS (the source for a product structured WBS) in a manner that prevents unnecessary splitting of safety critical functions that would complicate interface control and team understanding of adverse couplings.

- Defining the necessary functions allows a clear understanding or statement of the problem to help guide and define appropriate solutions. Often a solution becomes evident after a clear statement of the problem.

Step 3 Make it work. Create the simplest conceptual design of the contemplated system.

- Start with the simplest, most robust, and highest performance design option as the primary leg for accomplishing the mission functions identified above with inherent safety. The primary leg also forms the first leg when assessing fault tolerance. The simplest solution should lie within the constraint box boundaries with adequate margin for the succeeding steps below.

- If the simplest solution falls outside of the constraint box, then there may not be a workable solution; start the process over again with an alternate set of needs, objectives, and constraints.

- If the solution falls inside the constraint box but is not viewed as viable or optimum, consider alternatives with different operational concepts, designs, or derived requirements as depicted in the iterative loop shown in Figure i-4.

Step 4 Make it safe. Add diverse or independent elements to the simple system of step 3 that operates at lower or even marginal performance but with higher reliability as necessary to meet safety needs. This additional leg adds to system fault tolerance, although it may be applied as the last leg not necessarily the second leg. A simple diverse system maximizes the independence from prime system faults and should be easier to understand and verify.

- Evaluate the conceptual design and operations concept to determine potential failure modes and safety impacts. Initially the evaluation must be performed from the top-down starting from the mission level and consider each operational phase or operational system configuration of the mission. Utilize Functional FMEAs (based on functions) and /or fault tree analysis (top-down based on undesired consequences) along with an integrating technique such as ESDs to identify risk drivers.

- Utilize risk and reliability modeling techniques to bound the likelihood of the identified safety drivers. Discussions and debates resulting from likelihood and consequence discussions are helpful for further understanding and exploring the risk drivers.

- Pay particular attention to common cause failures that may defeat the intended safety improvements of the additional elements.

- Iterate the candidate mission rules and procedures to safely achieve the minimum acceptable objective.

- Provide an abort mode for those phases of the mission where the likelihood or consequence of safety critical initiating events or consequences cannot be contained.

- An effective methodology is to start with the "end game" of returning the crew to Earth and continue to work backwards from re-entry to launch assuring that safety and reliability are preserved during each operational phase of the mission. In other words, utilize technical resources such as, mass, volume, power, etc., to get the crew home first.

- If the solution is not safe, consider alternatives with different operational concepts, designs, or derived requirements as depicted in the iterative loop shown in Figure i-4.

Step 5 Make it reliable. Consider additional elements or other "legs," preferentially an additional primary leg of equivalent performance but not necessarily identical design for mission success. Additional legs for mission continuance add to system fault tolerance. Determine if the addition of the mission success leg leads to a safer system by considering all the potential dependencies.

- Utilize risk and reliability modeling techniques to estimate the effects of one alternative over another. If an alternative reduces overall risk and is affordable, add it; if not, be sure that the implications of accepting this risk are understood. Again, discussions and debates resulting from likelihood and consequence discussions are helpful for further understanding and exploring the risk drivers.

- Pay particular attention to common cause failures that may defeat the intended safety and reliability improvements of additional elements. Strive for designs that will limit the occurrence or consequence of common cause failures.

- Consider the maturity and complexity of the system when addressing how to mitigate unknown unknowns. This may drive additional features to facilitate testing and verification needs, for example additional test points or data recorders.

- If the solution is not reliable, consider alternatives with different operational concepts, designs, or derived requirements, as depicted in the iterative loop shown in Figure i-4.

Step 6 Make it Affordable. Estimate cost and schedule to develop, produce, and operate the system design of steps 2 through 4.

- Upfront design work has a high degree of leverage on the system's cost since these early activities expend around 10 to 15 percent of the project cost yet commit in excess of 50 percent of the total run out costs.

- Iterate the operations concept, design, or derived requirements as necessary to satisfy constraints, go back through steps 2, 3, 4, and/or 5, as necessary.

Step 7 Capture the Conceptual Design.

- Capture the decisions of steps 2, 3, 4, and 5 as the derived requirements, baseline operations concept, and baseline conceptual design.

- Consider all the legs of the system design when assessing system fault tolerance utilizing the rationale developed in the above steps to justify any differences between the selected approach and the starting point of two fault tolerance.

- Capture the allocation and utilization of technical resources along with the rationale for the allocations. (mass, volume, power, fuel, etc)

- Develop a program plan that tentatively defines prioritized requirements for each system element in the PBS, allocates physical and resource constraints to each, describes a system acquisition strategy, and assigns management responsibility for each effort. At the completion of this step, a safe and reliable system, producible at minimum cost, schedule, and complexity has been identified.

The previous steps describe a methodology for developing a design that can meet the needs and operational requirements through an iterative loop until performance, cost/schedule, and risk constraints are met. Even though the process is described in a step-by-step fashion, aspects of performance, safety and reliability, and affordability are not independent quantities and should not be considered independently in the process of design. While safety is of paramount importance, the implied order or hierarchy to the design process obligates the designer to make the design work first, make it safe and reliable, and then assure it is affordable. This is because affordability is moot if the design will not achieve a reasonable level of safety and reliability; safety and reliability are moot if the design does not function.

Utilizing this iterative loop described above and shown in Figure 2.3-2 provides affirmative rationale for the system design, its complexity, and the existence of each system element. This approach may lessen the likelihood of having to lop off pieces of a design to get it back "in the box." Lopping invariably leaves the system in a less cohesive state, vulnerable to unexpected interactions and other shortfalls.

ESD Example

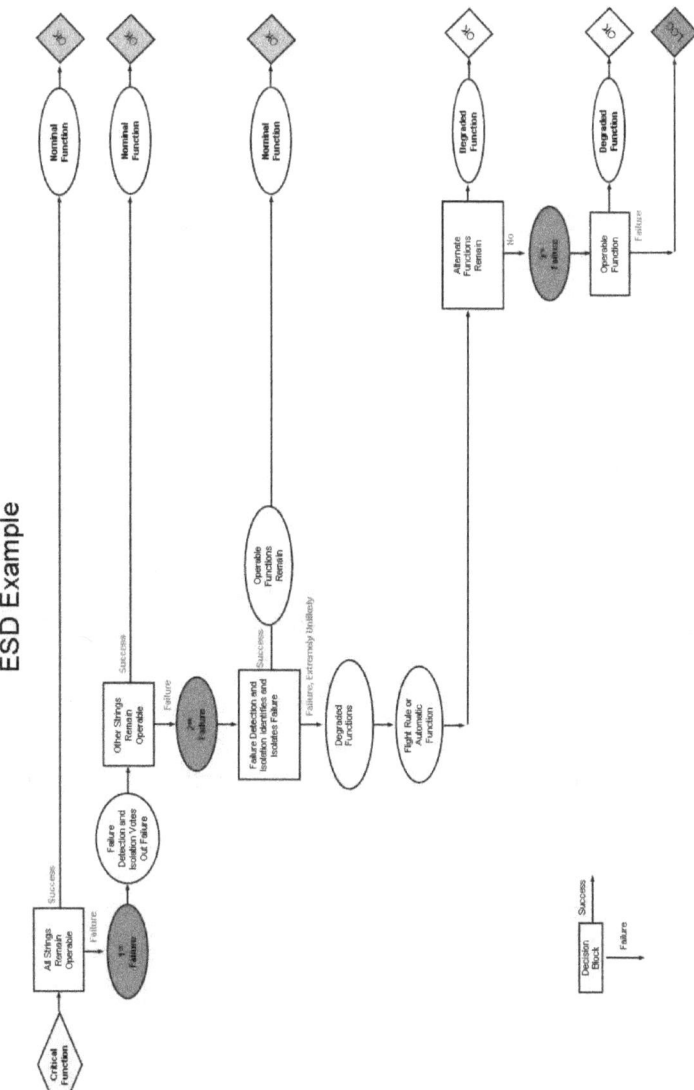

Figure 2.3-3 Event Sequence Diagram Example

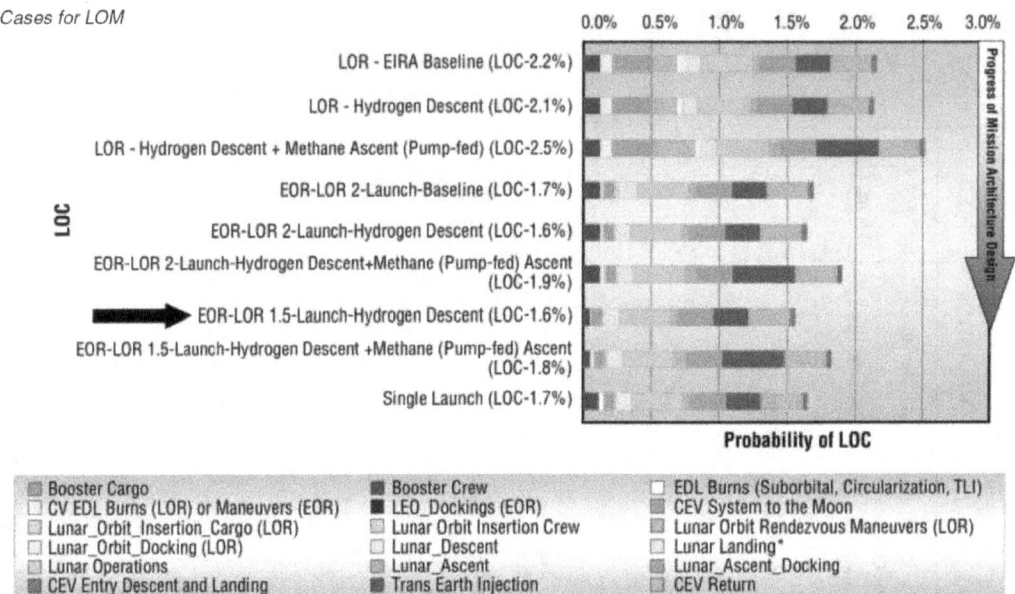

Figure 2.3-4 ESAS Variation of Mission Architectures with LOC Risk[12]

---

[12] NASA's Exploration Systems Architectural Study, Chapter 8 Risk and Reliability, Final Report, TM-2005-214062 November 2005

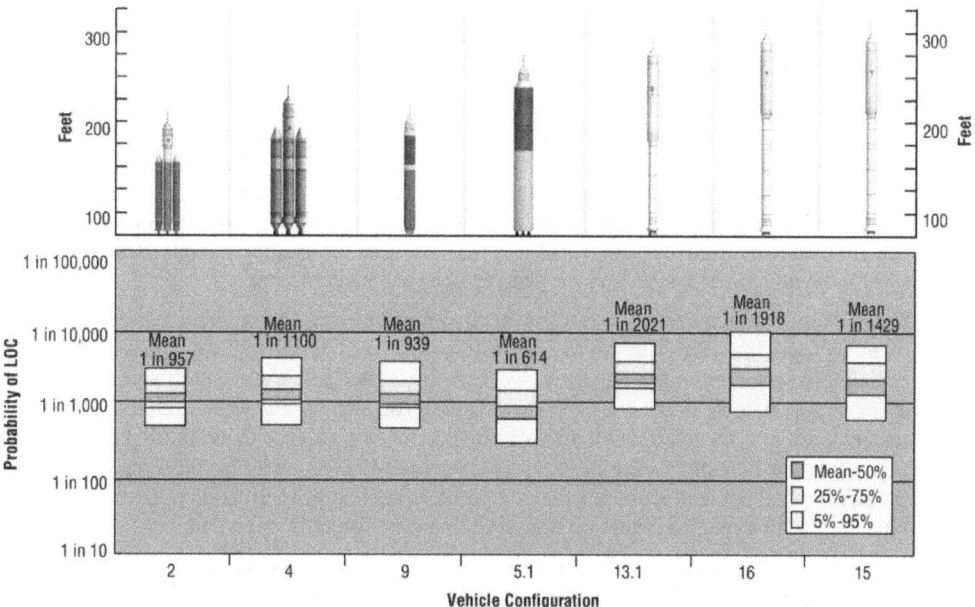

Figure 2.3-5 ESAS Probability of LOC Variation with CLV LEO Configuration [13]

2.3.4 System Redundancy Design Guidance or Rules of Thumb

Here are some rules of thumb for making risk trades for systems. These should not be considered "rules" as they are broken often. They should be considered as areas of concern and guidelines when considering alternatives, making design decisions, and when listening to arguments.

1. Simple is better than complex.
2. Passive is better than active.
3. Active two state devices are better than continually acting devices, e.g., boundary valves that just open and close are better than control valves that have to move to and hold a commanded position.
4. Passive devices are better when subjected to well-understood load spectra, than when subjected to unknown or uncertain loads.

---

[13] NASA's Exploration Systems Architectural Study, Chapter 8, Risk and Reliability, Final Report, TM-2005-214062 November 2005

5. Continually acting devices that move over a range either, linear or rotational, such as valves, are better than continually rotating devices such as pumps and turbines. In the latter case, lower rotational speed is preferred over higher.

6. Electrical circuits should only be given the bandwidth necessary to perform the desired function to limit the potential adverse and cascading effects of glitches, transients, Single Event Effects, etc. This is especially important with the advancement of high speed, low voltage, and low power CMOS integrated circuit, ASIC, and FPGA technologies that are becoming more sensitive to low levels of noise. This principle is important for devices or functions that are ultimately activated by pulses such as pyros, thruster valves, separation systems, deployment systems, reset signals, electrical switches, etc.

7. When safety enhancement of a simple system is required, robustness is preferred over diversity, and diversity is preferred over duplicative redundancy. This depends on the types of unknowns being protected against. This can be related to the concept of simple (simple systems may operate in complex environments). Robustness is always limited by understanding of the loads and materials. In some cases, defense in depth (diversity) better protects against unanticipated failures and failure modes, but both provide a measure of protection that duplicative redundancy does not. This is especially good if the diversity emanates from a required function (i.e., RCS back-up for main engine burns).

8. When identical unit redundancy is employed, there is always "coverage" or "common cause" or "correlated failure" factors to be considered that diminishes the theoretical benefit provided by this type of redundancy. This coverage is non-linear as more units are added, so that at some point the addition of another unit is counterproductive. In addition, the coverage effect is non-linear with the reliability of the individual unit. As the individual units become more reliable, the unreliability in the interconnections overwhelms the benefit of adding units.

9. For fast acting or "pulsed" service, such as is the case for combustion injection, solenoid valves or equivalent are to be preferred. For slower acting or boundary, that is leakage prevention or isolation service, motor operated valves are preferred.

2.3.5 Lessons from the real world[14]:

- Objectives and constraints are sacred. Therefore, they must be selected with care.
- Requirements, like ICDs, are tools used to ensure coordination and the accomplishment of objectives. Unlike objectives, they may be adjusted as required to achieve the objective within constraints. Each requirement has a direct and indirect cost. Accordingly, the total cost of satisfying each requirement must be individually weighed against its contribution to meeting the objective. Once accepted in step 5, requirements may be eliminated but

---

[14] Assembled for this report by TK Mattingly

experience shows such action seldom recovers schedule and may even increase cost. Therefore, requirements must be grudgingly accepted in step 5.

- The design and operational concepts must be jointly developed and set the minimum cost and schedule. Therefore, the engineers and operators are the only ones who can ensure the achievement of program objectives within constraints.
- The most egregious flaws in program estimates result from errors of omission.
- The iterations performed during steps 2 through 6 are most effective when conducted by relatively small co-located teams of personnel who will be responsible for execution.
- The design of integrated space systems is so complex that very few design principles can be applied without careful consideration on a case-by-case basis.
- Design and operational choices in human spaceflight must treat the consequences of component failure as more important than the probability of success.
- Use of heritage designs can be cost effective but require the same level of detail understanding as demanded of new designs.
- History must be considered but only to the degree it is directly applicable to the extant situation. There is a real danger that blind acceptance of yesterday's lessons will be inappropriate if the time between the lesson learned and the application to a new design is significant. For example, during development of the STS, the astronaut office insisted that critical displays and controls be implemented as conventional electro-mechanical designs due to the history of frequent failures in new electronic interfaces. By the time the Shuttle flew, electronic D&C had matured to the point that their MTBF was vastly superior to those demonstrated by the highest quality electro-mechanical devices. In fact, a major upgrade to the Shuttle was its incorporation of a glass cockpit, a capability that could have been built in from the beginning.
- PRA techniques are useful in evaluating relative merits of alternate design implementations but cannot be relied upon to predict actual reliability. This is especially true for devices with a relatively small population.
- As the design matures, one should expect to see an increase in weight, complexity and cost.
- The complexity and uniqueness of individual spacecraft designs require that special care be taken in accepting lessons learned on "similar" designs and applications without verifying the assumptions, purpose and basis for the presentation of results.
- Few real programs have the luxury of thoroughly completing the front end work, steps 1 and 2, before moving into step 7 and program implementation.
- There is usually more than one system implementation that will satisfy the minimum set of functional requirements. Among those that provide adequate performance, the selection should be based on optimizing other program objectives such as fault tolerance, ability to gracefully accommodate uncertainty, operational simplicity, cost or schedule. For example,

STS selected the aerodynamic FCS that provided adequate controllability with the greatest tolerance for aerodynamic uncertainties.

2.4 Implementing the System Right, Achieving a Safe and Reliable System

After the architecture is defined and the requirements are documented, the selected concept must be matured into hardware, software, logistics, etc., through the "Preliminary and Detailed Design," "Manufacturing and Assembly," "System Integration and Verification," and "Operations" processes shown as the major blocks in Figure i-1. These activities, which are performed throughout the life cycle, are based on well-established practices. Practices captured in established NASA and military standards will not be repeated in this report. The following sections will describe activities and practices focused towards implementing a reliable system and screening for potential problems.

Behind the life cycle activities are processes that form a multilayered approach for developing a safe and reliable system as shown in Figure i-1. Teams should view the processes of design, manufacturing, independent review, inspection, and test as multilayered activities assuring the implementation of a safe and reliable system.

Proper implementation of proven practices and processes in all the layers greatly improves the likelihood of success. The virtual series of nets (Figure 2.4-1) seek to prevent a potential hazard from resulting in a failure or a mishap thereby ensuring the system works and is safe and reliable. Each layer provides the opportunity for developers to identify differences between the designers' intent and reality allowing corrective action before the system is implemented, thus precluding failures. Thoroughness and strength of the layers depend on critical characteristics of team members (poles in graphic). Each layer within the multilayered approach provides a mechanism for identifying (bracketed items in the graphic) and collecting off-nominal conditions, warning signs, and precursors to failure. Conditions that could result in significant risks should be input and integrated into the risk management process as described in Section 2.5.

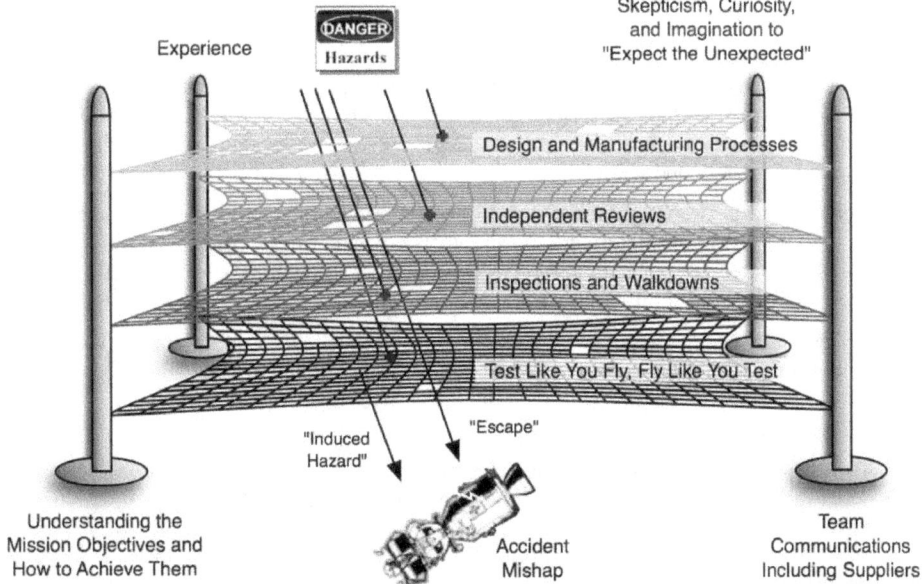

Figure 2.4-1 Multilayered Approach to Produce a Safe/Reliable System and Screen for Hazards (Adapted from James Reason[15])

- Dense and diverse nets with solid supporting poles serve as barriers, or screens, preventing hazards from causing accidents or mishaps
- Multiple imperfections in the nets, or supporting poles, may allow hazards to result in accidents or mishaps
- Avoid inducing hazards or latent failures into sensitive system elements by processing steps

The challenge for the development team is to produce and operate a system in a manner that maximizes the chance of choosing the right solution and uncovering problems before they cause significant adverse consequences. Given finite resources, it is the designers' challenge to commit technical, cost, and schedule resources in a way to achieve a desired level of reliability. Strengthening weak layers, through proper implementation of design choices, can be viewed as investments in improving reliability by decreasing the likelihood of flaws and increasing the

---

[15] Reason, J. (1990) Human Error. Cambridge: University Press, Cambridge
Reason, J. (1997) Managing the Risk of Organizational Accidents. Aldershot: Ashgate.

likelihood of detection. With judicious investments, the probability of failure can be made extremely remote.

To do this, the designer must be aware of the types of flaws that can occur, processes that introduce flaws, and ability of the different layers to prevent failure. When the source and character of flaws are understood, then targeted barriers can be used to detect and prevent potential failures. Conversely, uncertainty in the characterization of flaws leads to the need for more diverse flaw detection methods.

The thoroughness and strength of the layers depend on critical characteristics of team members. As in Figure 2.4-2, these characteristics are shown as poles that support the nets. As described in Section 2.1.4, it takes experience to produce a reliable system and curiosity, skepticism, and imagination to foresee what might go wrong, and then act proactively.

No single layer can assure reliability by itself. Producing a reliable system requires the proper application of process at each layer. By the same token, no individual layer should be compromised based on the assumption or hope that another layer will compensate for deficiencies of another layer.

The following sections will describe each of the design, manufacturing (fabrication), independent review, inspection, test, and operation layers. How each layer helps to identify potential system weaknesses and contributes a safe and reliable system is described.

2.4.1 Design

Designing a system that meets all requirements within mission constraints is an exceptionally difficult process requiring experience and skill. Like the architecting process before it, the design process also utilizes an iterative approach to achieve an optimum result. Results of the previous phase, including baseline architecture, operations concept, and system requirements are the initial inputs. Candidate designs are proposed and assessed to examine whether mission level requirements are met within program constraints of cost, schedule, performance, and risk (refer to Figure 2.4.2). This is accomplished by flowing mission-level requirements down to lower levels of the design, elements, subsystems, components, even parts and, concurrently, into disciplines. Each requirement described should be necessary, stated unambiguously, and verifiable.

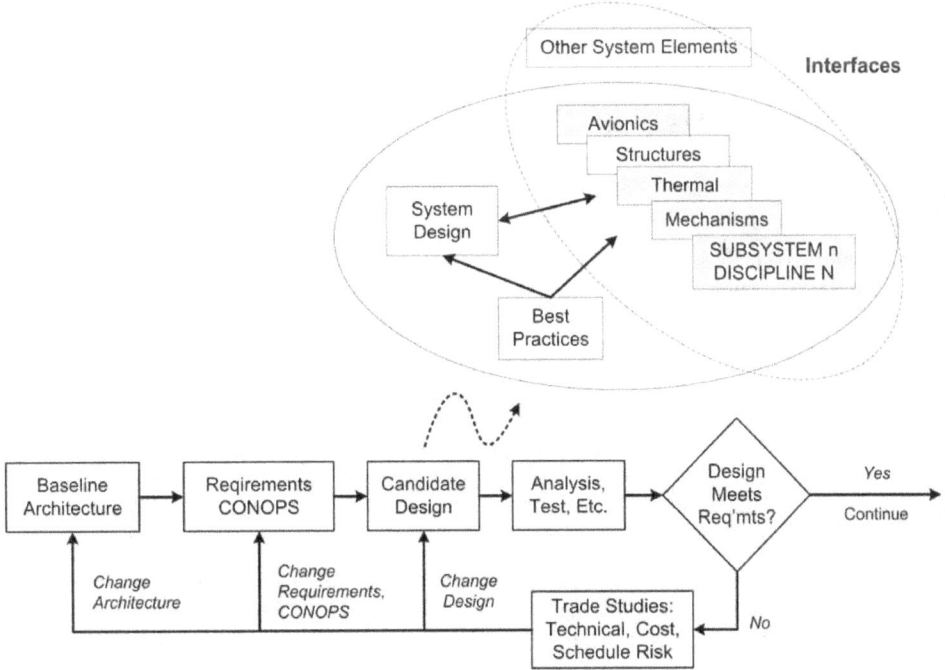

Figure 2.4.2 Candidate Design and Assessment Iterative Detailed Design Loop

To achieve a reliable design, the iterative flow, shown above, should utilize risk-based evaluation criteria to identify risks to safety and mission success requirements. To meet requirements, alternatives are considered and decisions are made as to whether to change the design, suggest alternate requirements, or change the operations concept. Reliability analysis helps identify design weak points and identify alternatives. By applying a risk-based approach as discussed in Section 2.3, designers carefully examine where resources should be expended to refine spacecraft designs to meet safety and mission success requirements. Reliability analyses are an important part of this process, designing fault avoidance and tolerance into the system from the "ground up."

Reliability analyses are used to ensure implementation approaches are sound and consistent with project requirements and the intent of the architecture. Refer to Section 3.0 for details. Results from reliability analyses may drive design considerations to reduce failure likelihood, mitigate risk, increase margins, or implement contingency plans. Special steps are often documented in the Critical Items Lists or the Hazard Control portion of Hazard Reports to assure the system element is designed and produced with the requisite attention to detail.

Some important considerations in developing reliable designs are listed below.

- Design must be testable, manufacture-able, and operable. Specialists from these disciplines should be involved as early as possible.
- The entire operational life cycle should be considered, including flight and ground operations, maintenance, and refurbishment, if applicable.
- Knowledge of a system's sensitivities along with uncertainties in environments and system parameters forms the basis for defining margins. System sensitivities should be assessed and should be minimized where possible.
- Margins should be included that are sufficient to cover the entire range of possible inputs (such as environments). A method of including robustness is to set requirements over expected levels (expected + margin).
- System designers need to recognize that heritage designs or COTS products do only what its initial builders intended, not necessarily exactly what is needed in the new application. Use of heritage hardware (qualified by similarity) should be applied with caution. Often, it takes only a small deviation from a previous design or application to produce vastly different behavior. "Generally throughout the agency, heritage hardware (and software) are given less scrutiny than new hardware because of an often unfounded faith that heritage designs are qualified and will perform properly in a new application."[16]
- The reuse of existing designs, heritage elements, and COTS hardware and software should undergo extensive scrutiny to verify that the item meets all requirements for the given application. Great care must be taken to assure that the previous work applies to the mission at hand.
- Utilization of existing designs, heritage elements, and COTS hardware and software requires the consideration and accommodation of its interfaces and constraints potentially driving other aspects of the system. Reusing existing designs means accepting the way the product works and the way it interfaces and interacts with the rest of the system. Designers may need to adjust the system to accommodate the constraints of the reused elements.
- Beware of the domino effect where fault tolerant architecture is not a preventative or solution to a system design configured using low reliability components (heritage or COTS selection based often rationalized on low cost).
- Consider potential spatial interactions and identify where spatial separation is necessary to provide the intended fault tolerance. Physical placement can subject components to coincident hazards (e.g., thermal, vibration) or common cause failures that can cut through barriers intended and built into the design.
- Crewed vehicles are often designed to have portions reused from flight to flight. Plans for refurbishment should be considered at the outset of a design and refined over time as the

---

[16] Genesis: Mishap Investigation Board Report, Volume I, November 30, 2005.

program progresses. Once returned for refurbishment, hardware should be thoroughly inspected, to predetermined criteria, for nominal or off-nominal issues warranting further investigation for insertion into the risk assessment/corrective action process. Hardware discrepancies uncovered during this process should never be ignored.

Design Verification and Validation through Analysis, Modeling, and Simulation

Early in development, a verification matrix is generated to identify the method demonstrating how each requirement will be verified. Modeling, simulation, and analysis are frequently used during design to not only verify and validate a design but also to approximate the system's performance and predict its ability to withstand operational environments. Analysis is frequently used to verify parameters that are outside the ability to verify by test. It is important to use correlated models and simulations to estimate real-world behavior. Analyses and models not validated to real-world results or correlated to physics could introduce risk by under or over estimating real-world effects. Designers should plan development testing as early as possible to verify both models and simulations. Other verification methods, tests, and inspections, will be discussed in subsequent sections.

Design's Technical Integration

As shown in Figure 2.4-2, subsystem teams need to integrate their elements, and understand how they influence other subsystems and how they are influenced by other subsystems. It is important for the SE team to recognize the importance of fostering interaction among discipline engineers after requirements have been allocated. Extensive communication among the teams developing the design pieces is essential to achieving a balanced design. Both formal and informal communication is required. Teams should be aware of unwanted interactions among subsystem elements and consider the potential effects of hardware and software anomalies. Interfaces must be considered during this process by relevant cross-interface information contained in an ICD.

The interactions and influences among systems are important as engineers perform detailed design trades at their level until designated requirements (performance, environmental, etc.) are met. Resultant design pieces must then be re-evolved into a system-level design and reevaluated against system requirements. At the system level, adverse interactions will be found that could not be anticipated until detailed designs were developed at lower levels. Accounting for, or making adjustments to, these adverse interactions early in the process enables the item being developed to function properly at the system level; assuring a successful system design.

Another important aspect of technical integration is the revalidation of the design as it matures. The early systems analysis activities described in Section 2.3 rely on simplifying assumptions, linear modeling, and simple interfaces. As the design matures and discipline experts proceed with their detailed design, the physical non-linear realities start to surface and may add complexity and unintentional interactions. SE needs to revalidate the system design, compatibility of interfaces and total integrated system performance as the design matures. A thorough discussion of this process and considerations for integrating various subsystems into a cohesive system is found in NASA TP-2001-210992.

Assessing Compliance

After design iterations and analysis are complete, some parts of a design may not meet their requirements due to various other constraining factors influencing the overall design. In these cases, design waivers or deviations may be necessary to resolve the noncompliances. In cases where waivers or deviations increase safety or mission success risk, these items need to be input to the risk management process for total project risk assessment as discussed in Section 2.5.

2.4.2 Manufacturing, Assembly, and Integration

Spacecraft and component manufacturing occurs at a number of geographically diverse sites, including NASA, prime contractor, subcontractor, and manufacturer facilities. This large base of suppliers alone provides the rationale for why proven manufacturing techniques should be consistently applied to all system elements. No supplier should unknowingly introduce a weak link into the total system. Manufacturers need to meet certain standards to produce reliable hardware. Projects seeking to produce reliable hardware need to establish manufacturing, workmanship, and inspection standards to ensure consistent quality and reliability of their system elements. Process controls should be instituted such that personnel are trained to implement the process, understand deficiencies when they arise, correct those deficiencies, and retrain the modified process.

International process guidelines like ISO 9000 help with methodologies for implementing this approach, but do not define the actual manufacturing processes ensuring consistent quality among all suppliers. ISO 9000 is only a portion of quality control for a successful manufacturing process. Equal care must be exercised in process development and implementation.

The use of heritage hardware and COTS products presents unique challenges for reliable systems. COTS suppliers offer limited visibility into how their products are produced. Gaining insight into COTS elements may be necessary if they are used in mission critical applications. Even reputable COTS vendors can produce products that have defects. COTS vendors are not generally constrained to maintain the same design, production process, material, or components over time; only to produce a product that meets the unit level specifications. There is the potential for significant lot-to-lot variability. Care must be taken to ensure the unit flown or used in support of flight is the same as the one inspected or tested. Teams need to identify those items that warrant additional insight and implement additional controls, as necessary, to assure safety and reliability.

An environment of freedom to report problems without fear of reprisal, where safety and mission success is the primary goal, is essential. If, during the course of manufacture, an item is discovered not to meet drawing or process requirements, a flag should be raised. In response to the flag, the team must choose the appropriate resolution and disposition considering safety and mission success. If waivers, deviations, or Material Review Boards (MRB) are processed, they should only be approved after a complete assessment demonstrating acceptability from a risk perspective. Waivers should be viewed as a last resort, not as a routine practice to save cost and schedule. See Section 2.5 for discussions on integrating risks and Section 12.0 for more information on the Materials and Processes aspects of manufacturing.

Fabrication includes both construction of early prototypes used for development testing and the building and integration of the flight article. It should be noted that prototypes are not always built to the same level of detail or quality as flight hardware; therefore care should be exercised that they replicate component equipment and performance as closely as possible. This is particularly important where the intent is to retire flight article risks. A qualification article, if used, must be built to the same standards as flight hardware. Liberal use of these early pathfinders, especially for new technology, as a method of uncovering design flaws while they are still relatively inexpensive to correct is highly encouraged.

Software is becoming increasingly important in the successful functioning of space systems. As reliance on software for critical functions grows, so does its contribution to mission failures. Flight software is developed iteratively and in pieces, similarly to flight hardware. Configuration control, to track changes and versions, as well as ensure they do not detrimentally affect other part of the system is important, as is verification that is becoming increasingly difficult. See Section 6.0 for detailed discussions about Software.

Integration is the point in the systems development process where all of the elements of the system come together. Portions of the system may have been build in various and sometimes remote locations, in some cases elsewhere in the world, before they are delivered to the integration site and before they are combined into a single complex functioning system. Unexpected occurrences often arise, where pieces of the system do not work together as they should, either mechanically or electrically. No integration anomaly should be left unaddressed. Proximate and root cause for each problem should be found, risks identified and fully understood, and corrective action determined.

2.4.3 Independent Review

Independent review along the development and operations life cycle provides a critical layer in producing a reliable system. Reviews allow the access to knowledge and experience from outside the development group. Independent peers can identify issues or suggest alternatives that may not be known to the developers. As discussed below, engineering peer reviews can provide valuable input to the development team to make unknowns known.

Independent means review team leaders and team members who are not in the direct chain of command or the program or project.

A review is an evaluation of the system elements by a knowledgeable group, independent of the project or element under review, for the purposes listed below.

- Identifying potential problems, unexpected interactions, and/or risks and recommending improvements and corrective actions
- Timely identification of deficiencies that will allow the Project to take appropriate and efficient corrective actions
- Validation of the adequacy of assumptions, trade-offs explored, and design solution proposed
- Assessing the status of and progress toward accomplishing the planned activities
- Making judgments on the activity's readiness for the follow-on events

Reviews are conducted consistent with maturity as defined by life cycle phase and expectations established by review success criteria.

Space system acquisitions utilize a series of formal program reviews (System Requirement Review, PDR, Critical Design Review, etc.) as milestones at which the Project is expected to have achieved a prescribed level of accomplishment. The NASA Systems Engineering Handbook SP-6105 and others such as MIL-STD-1521B, describe these reviews in detail. While these reviews are valuable in providing a point at which the Program can be thoroughly assessed (often forcing self-assessment), they are frequently misused. To have validity, the reviews must be gates through which the Project is not permitted to pass without the required progress toward a low-risk design. However, cost and schedule pressures have often resulted in continuation through these gates prematurely, ironically increasing risk rather than reducing it. In addition, issues raised by reviewers on formal Review Item Discrepancy (RID) forms in some cases have been ignored precluding early resolution to impending problems.

Timely closure, RIDs, Requests for Actions (RFAs), and Action Items submitted at a review are critical to risk reduction. The review process provides these instruments in an attempt to document risks. Some RIDs or RFAs that remain open or may not be closed and agreed to by the initiator could represent risk to the project. These open or unresolved items need to be input to the risk management process for further assessment.

Among the most value added reviews are engineering peer reviews (EPR) that are convened when an in-depth penetration by independent subject matter experts (center personnel and non-center personnel) is needed to scrutinize details of the element under review. Large formal reviews provide some of the benefits of EPRs but primarily provide insight at a higher level where delving into details is difficult. Higher-level reviews typically look into the project status and issues being worked in a comprehensive framework at measured milestones.

Organizations define the EPR standards and guidelines that apply to the EPRs they conduct in support of flight projects, both for work completed in-house and at contractor sites. Review benefits depend strongly on the quality of the reviewers and follow-through on identified issues, concerns, and recommendations. Planned contractor participation in NASA reviews and NASA participation in review of contractor-provided product(s) must be documented in the Project

Review Plan and made known to contractors in procurement documents prior to contract negotiation.

2.4.3.1 High Value EPRs:

1. EPRs need strong experienced leaders, broad systems people, as well as detailed specialists, for each review. Level and experience of peer reviewers is inversely proportional to the experience of the designer.

2. EPRs need to be properly planned and scheduled to lead up to more formal reviews. There can be hundreds of peer reviews on a mission.

3. Peer reviews are the only place where the detailed drawings and detailed analyses that supports the design can be discussed.

4. It is important for designers to discuss functional/operational behavior, not just how the design meets requirements. The process of discussing expected performance provides opportunities for insight and unexpected information and can fill gaps in requirements or designer's experience (e.g., including the right filtering in noisy input circuits).

5. Use of heritage hardware and software should undergo a "Heritage Review" to assure that assumptions, environments, and operational sequences allow the product to fulfill its intended need in the new application. Ensuring existing, inherited, and COTS designs require the same rigor in peer review as new designs. Making assumptions about inheritance that bypasses the peer review process can lead to serious problems later.

6. Peer review allows some level of penetration of culturally/experience-based approaches that could be blind to a new environment. Examples: 1) flawed assumptions can be made about implementing fault protection for systems in near-constant communications with the ground versus those with time delay or significant periods/events requiring autonomy; and 2) Cassini Huygens probe radio design did not have the right Doppler shift included.

7. Ground support equipment (GSE) interfaces to flight systems need a thorough peer review, including failure modes and effects (e.g., Deep Impact failed to detect a GSE design flaw where the failure of a GSE switch would cause the failure of a flight power switch).

8. Peer review allows communication about known areas where problems can/have occurred in the past that may change with technology, e.g., asynchronous versus synchronous gate array designs.

9. Peer review should be applied to test planning and results to ensure verification and validation process catches anything that the peer review process misses.

10. Incorporation of diverse design teams from multiple NASA centers provides alternate perspectives of design elements and challenge group think.

2.4.4 Inspection and Walkdown

The inspection and walkdown layer encourages teams to look at the product incrementally as it is produced as well as when it is assembled in its final state. Requirements, drawings, and processes intend to prescribe how a product should be built. To assure that the product was actually produced properly per the paper and designer's intent requires teams to look at the product, measure it, and verify that it was built with the intended quality.

Inspection is generally used to assess the satisfaction of quality control requirements. Valuable information on a product can be gained from inspection alone. Inspection should not, however, be used to verify complex functional or environmental performance. System reliability requires proper operation over time, and the time aspect cannot be inspected. As such, quality control inspection is necessary but not sufficient for ensuring reliability.

Inspection is a verification method of examining the product itself, using visual means or measurements requiring precision measurement equipment. Inspections can be performed by independent "Quality Inspectors" or by engineers familiar with the design depending on the purpose of the inspection. Walkdowns are performed by teams to ensure the product was produced as intended.

Certain components with safety and mission critical functions may require special in-process inspection points to assure their quality. These critical inspection points should be planned and inserted into the process. Sometimes it is necessary for design engineers and/or the users, such as the crew, to look at the system. Often inspection points are required by established manufacturing standards. Special inspection points may also be identified in reliability analyses, failure retention rationale such as those documented in the critical items list (CIL), Hazard Report Controls, or risk mitigation plans.

It is critical for inspectors to have the appropriate experience and training. It requires significant experience and/or training to understand what is critical for reliability and know what types of things to look for.

Once a discrepancy or issue is identified through the inspection process, teams need to decide how to resolve the issue. Discrepancies can serve as warning signs or precursors to failure. Often issues are corrected. When discrepancies cannot be corrected or made compliant, then teams may take on risk. Inspection discrepancies or MRB actions with safety or mission success impacts may represent risk and should be input to the risk management process.

2.4.5 Product Verification and Validation, "Test Like You Fly"

After years of work and thousands of pragmatic compromises, the system that is to be flown will inevitably be different from that envisioned in many subtle ways. As the hardware and software coalesce, the opportunity to make system changes essentially vanishes, although the mission objective remains unchanged. At this point, program emphasis shifts from ensuring that the elements have been produced per specification to assuring that the system's functional capabilities and limitations are thoroughly understood and accommodated by the flight plan.

Testing provides the most effective method of proving that the system behaves as it must according to requirements (verification) and as it should according to the user needs (validation). Testing represents a critical layer in the multilayered approach for assuring a safe and reliable system. The axiom is "Test like you fly; fly like you test." Tests should replicate actual conditions to the maximum extent possible, including hardware, software, environments, interfaces, and operational sequences. Testing provides the opportunity to validate assumptions used during design and can uncover unexpected interaction among system elements and the users. Tests go beyond compliance to requirements and can demonstrate that the system accomplishes its intended purpose through mission simulations, end-to-end tests, and joint integrated simulations.

End-to-end system functional testing has historically been the technique of choice in satisfying the need to characterize the as-built system capabilities. During Apollo, this was accomplished through full duration thermal-vacuum testing of prototype spacecraft with crews executing near real-time mission profiles (except re-entry). This activity instilled high confidence at great cost, but the state-of-the-art simply could not support any other approach. By the time the Shuttle entered development, analytical techniques had matured enough that it was possible to trade off the high cost of building analytical models against the enormous cost of building and operating a thermal vacuum chamber large enough to accommodate a full size Orbiter. In so doing, reliance on the interaction of aerodynamic, thermal, and structural models with new systems became critical. As a result, the design of the Orbiter had to accommodate the planned in-flight verification of analytical models, an enormously expensive undertaking in itself. Although the Shuttle validated the efficacy of this approach, it also illustrated the need for extraordinary attention to detail and that the state-of-the-art in analysis has not achieved perfection. In fact, the implementation of this approach requires increased diligence in testing components to assure they have been functionally exposed to flight-like environments and operating sequences. It requires careful scrutiny of all test data and an understanding of the root cause for all unexpected test results. It makes the flight test program more, not less, important and requires the seamless integration of design, production, and operations disciplines throughout the development process. Finally, there is no substitute for end-to-end functional testing whether accomplished on the flight article itself or through a combination of laboratory and vehicle exercises.

Designing a system to be tested may well be the biggest challenge facing developers of new space systems.

To assure safety and reliability, functions that enable a system's fault tolerance and robustness through redundancy and backups need to be verified and validated. Each function intended to improve a system's reliability needs to be tested to assure that it works and that unintended interactions and operational complexities do not defeat its intended purpose. In addition to nominal testing, these tests can require the exploration of off-nominal conditions, off-nominal operational sequences, contingency procedures, and "negative testing."

Teams need to identify the parts of the system that cannot be tested like flight and ensure that alternative methods are sufficient and appropriate surrogates. Often testing is performed in pieces, via simulators, by analysis and/or a combination of approaches when a "Test like you fly" approach is not possible or practical. For these cases, great care must be taken to ensure the total system testing is valid.

Tracking of requirements verification, whether accomplished through tests or other means, is critical for system safety and reliability. Formal tracking systems should be used to assure each requirement is verified against the actual flight hardware and software and that verification results are peer reviewed.

MIL-STD-1540 can be used as a guideline for planning the test program that historically achieves success (refer to Section 1.2). Tests should be used to verify requirements are met *and* provide insight into actual system behavior, highlighting areas where response to test stimuli is outside of the expected or intended. Any areas highlighted are red flags of impending problems and should not be ignored, whether the letter of requirements statements are met or not. Tests should also be conducted using test margins to reveal behavior beyond nominal.

Tests are performed as increasing levels of assembly, from part to component through to system. A test at a lower level of assembly should not be used to justify omission of a system-level test since it might not satisfy a system level requirement. Conversely, a system level test should not be substituted for lower level tests. Box level tests, for example, may provide a more thorough understanding of component level behavior in the absence of external influences such that problems can be uncovered, understood, and corrected in a more timely and cost-effective manner.

Existing hardware/software, heritage elements, and COTS products can contain defects. All components of a system in mission critical applications; therefore, regardless of their origin, must be tested with the same rigor that is applied to new designs.

Flight software should be verified by using real-time simulations. Such simulations should run flight code on flight hardware components (hardware-in-the-loop). Crew interfaces should be incorporated by appropriate functions exercised by crewmembers or their proxy (with the participation of Human Factors Engineering).

Reviewing test results is as critical as performing the test in the first place. Tests seek to identify where the system may not perform as expected and that requires teams to review data, identify adverse trends, precursors to failure, and warning signs that the system may not function as intended. Teams should "listen to what the hardware is telling you."

When unexpected conditions or trends are observed they need to be documented and chased to proximate and root causes. Teams need to choose prospective corrective action with reliability, safety, and mission success in mind. Test discrepancies whose cause or corrective action cannot be definitively defined and their reoccurrence may have significant safety or mission success consequences should be flagged and input to the risk management process as discussed in Section 2.5.

2.4.6 Operate, "Fly Like You Test"

Although the conducting of mission operations is beyond the scope of this document, those operations do provide an important component of the process of designing for mission success.

Just as systems should "Test like you fly," systems must also "Fly like you test" to avoid encountering uncertain responses to untested conditions or sequences. That is, verifications only have validity if synchronized with the operations that they have verified. Thus, nominal and contingency operations should be conducted, as closely as possible, as planned in the previously conducted test program.

Despite all best efforts, anomalies will occur during mission operations. When they do occur, anomalies should be cataloged in a database using a rational structure that allows easy search for trends, repeating anomalies, and precursors to failure. Categorization of anomalies should be consistent with reliability, safety, and mission success drivers that may be used to uncover trends. Once understood, each anomaly should be chased to proximate and root causes. The cause is then translated into Program risk and, when risk is beyond acceptable thresholds, corrective action should be taken.

To satisfactorily resolve flight anomalies, sufficient data on vehicle performance must be available. This requires a set of flight instrumentation. The temptation to limit the instrumentation set to save weight, power, or cost should be resisted. Thorough resolution of any on-orbit anomaly helps not only the next mission of series, but has the potential to prevent failures on unrelated programs as well. Even if only unexpected behavior not resulting in a failure occurs, the data collected during the mission may be used to surface impending problems or improve models and simulations further reducing risk (refer to Section 2.2.2 Quantitative Requirement and Supporting Analysis, Establishing Failures Rates and Uncertainty).

Reviewing flight results is critical to identify adverse trends, precursors to failure, and warning signs that the system may not have functioned as intended. Systems often offer warning signs, "close calls," or "near misses" before failing catastrophically.

When unexpected conditions or trends are observed they need to be documented and chased to proximate and root causes. Teams need to choose prospective corrective action with reliability, safety, and mission success in mind. Flight discrepancies, whose cause or corrective action cannot be definitively defined and their reoccurrence may have significant safety or mission success consequences, should be flagged and input to the risk management process as discussed in Section 2.5.

2.5 Integrating Risk

Risk integration is an activity performed collectively by technical and management teams with the goal of understanding and appreciating the total risk state of the Program by consequence irrespective of the risk identification method. Risk integration provides management with information necessary to identify, understand, and evaluate risk trades throughout the life of the Program.

2.5.1 Identifying and Classifying Risks

One of the major challenges for Program-wide risk management involves integrating risks with differing consequence and rating scales into a single high, medium, low scale or a single 5 x 5 risk matrix. Disparate safety, mission success, and development risks are hard to compare on a single scale because of different consequence and likelihood scales as shown below. Safety, mission success, and development risk types utilize differing techniques for identifying individual risks as well.

- Safety risks involve personnel injury and loss of life; rating scales are logarithmic and range from 0.1 to 10^{-6} (NPR 8715.3 NASA Safety Manual).
- Mission success risks involve the inability to meet and complete the mission requirements; rating scales are also logarithmic with ranges from .1 to 10^{-4} (scale varies by program).
- Development risks involve the inability to deliver a quality product on time and within cost, often also called Programmatic Risk; rating scales tend to be linear from .1 to .9.

The objective is group risks with common or similar consequences together so that they can be evaluated and ranked. Figure 2.5-1 shows a hierarchy for collecting risks according to a consequence-based focus. The consequence focus helps team members focus on the top-down perspective while identifying risks and causes them to seek an understanding of the risk's ultimate consequence.

Figure 2.5-1 Consequence Focused Risk Types

Section 2.4 describes a multilayered approach applied throughout the system's life cycle to field a safe and reliable system. Inherent to each layer is a mechanism for capturing, evaluating, and resolving off-nominal conditions shown below:

- Design and manufacturing processes create <u>Waivers</u>, <u>Deviations</u>, and <u>MRB.</u>
- Independent technical reviews create <u>Action Items</u> and <u>RIDs.</u>
- Inspection and walk downs create in process <u>Inspection Discrepancy Reports.</u>
- Testing with a "Test like you fly" approach creates <u>Test Discrepancy Reports</u> or <u>Problem Failure Reports</u>.
- Identifying where "Test like you fly" cannot be followed and therefore accomplished in pieces can represent risk.
- Operating the system creates <u>In-Flight Anomaly Reports</u> and Trend Analysis Results.

Figure 2.5-2 shows a desirable risk information flow that includes the different risk identification methods that are used during all phases of a program. The figure includes typical risk identification activities, risk ranking, and summarizing activities that constitute risk integration. It also shows where the additional information on acceptable risk and project resources integrate into the risk management decisions. Decision outcomes are captured and utilized to reduce risk and communicate risk to stakeholders. Risks are continuously monitored to see that controls are effective and emerging or missing risks are identified.

Risk sources include off-nominal conditions and close calls that can warn of potential failures or serve as a precursor to failure. Each of these off-nominal conditions needs a closure process that identifies, captures, and integrates any residual risk, as well as a method to validate assumptions and models used in the risk analysis process. If risks cannot be definitively resolved, there may be a residual safety or mission success risk. These residual risks should be identified and tracked. Residual risk can be characterized by the potential for the problem to reoccur because the original off-nominal condition could not be either definitively identified or definitively corrected. An individual off-nominal issue may not represent a risk by itself, but a large number of individual small risk issues can sum and integrate to a large amount of total program risk.

Warning signs and potential precursors to failure evident from trend analysis, "close calls" or "near misses," provide useful inputs for risk managers. Trend analysis, using tools such as SPC or Crow-AMSAA, can provide advance warning that performance or margins are degrading. "close calls" and "near misses" indicate the reduction of margin or indicate the potential for more serious consequences should the operational sequence or the environment vary in an adverse direction. Some of these unexpected situations may indicate that assumptions or modeling needs to be updated. Teams should establish a mechanism for identifying these kinds of warning signs or precursors and incorporate results into risk assessments.

Total program risk is represented by the accumulated and combined risk of all these sources described above. There are a number of tools and processes for managing risks, including continuous risk management (CRM), NASA NPR 8000.4[17], and MIL STD 882. These processes require continuing activities to identify, analyze plan, track/monitor, and control risks as they are identified and ideally retired during the program. CRM is being implemented throughout NASA, and including Exploration Systems Mission Directorate. It is very important the entire program be vigilant in identifying and closing out risks.

The risk management process is something that requires participation rather than observation. Having NASA intimately involved in the design process, as part of integrated design teams that produce analyses consistent with the development of the design, provides real-time insight into system design development and allows NASA to collect information relevant to the system without the necessity of making special requests to the contractors or modifications to their contracts. Marshall Space Flight Center (MSFC) successfully applied this approach during the early development process of the TR-107 engine.

Indicators of a good CRM process are signs of management ownership during Program reviews and meetings (i.e., management is fully engaged in the process), the program is dynamic (i.e., risks are being identified and retired), and team members feel free to identify risks for the program.

---

[17] NASA NPR 8000.4, Risk Management Procedural Requirements w/Change 1

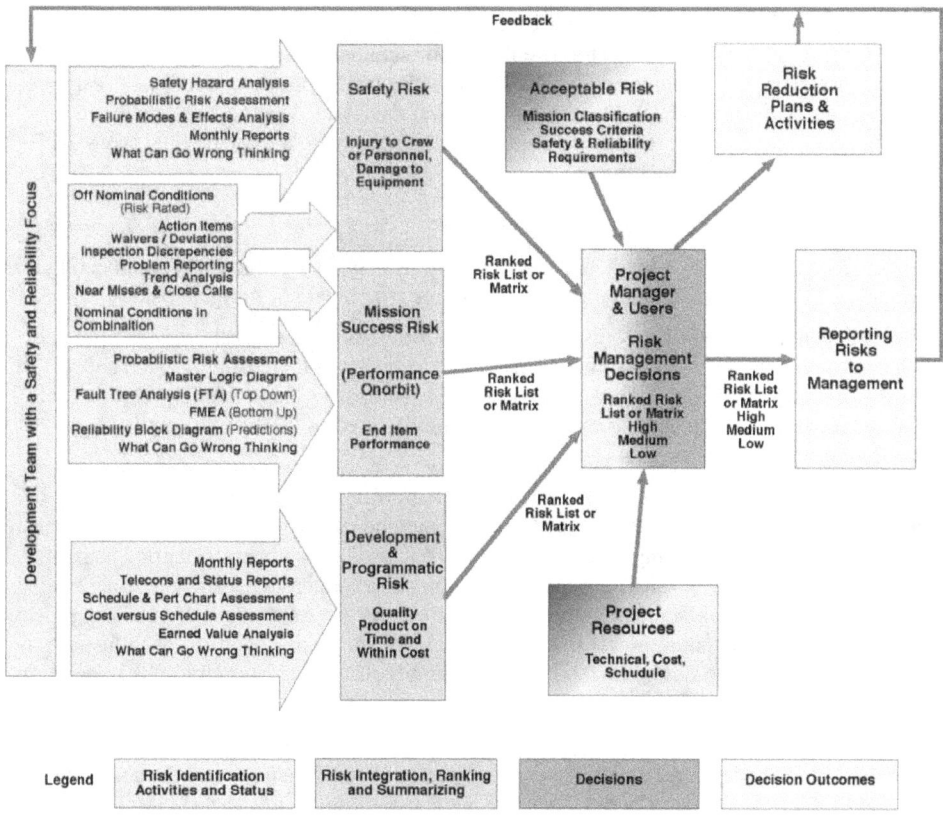

Figure 2.5-2 Risk Information Flow

CRM is in place throughout the life of the Program. SE interacts with the CRM process by evaluating the design requirements and constraints and identifying risks. Risks can be viewed as requirements unmet or constraints exceeded. The impact of the risk on the Program must be evaluated in a top-down manner across the Program to establish operating risk margin within the element. The likelihood of occurrence must also be evaluated by the Program element. This information is then used to prioritize efforts within Program elements to control risk within their purview. If a Program element risk becomes real (e.g., performance below/approaching minimum requirements, schedule delay, failure event), SE and risk models are used to respond to the problem from the top-down by changing requirements, or constraints, adding resources where most appropriate.

2.5.2 Incremental Acceptance of Risk

Sometimes risk is accepted and accumulated in small increments. Each of the increments by itself may not appreciably increase total risk. However, a large number of small risks can interact (amplify), aggregate, and accumulate to a much higher risk state. The tools and techniques should provide a mechanism for decision makers and managers to assess the total risk state represented by all the small increments of risk

2.5.3 Evaluating and Trading Disparate Risks

Comparing safety, mission success, and development risks types is important for allocating a common pool of technical, cost, and schedule resources to obviate or mitigate risk.

Trades between disparate risks require a technique for evaluating total risk. For example trade-offs between eliminating testing to reduce development risk, but increasing mission success risk requires a figure of merit in total risk space. Likewise, comparing the risk of hypergolic fuels versus the reliability of the propulsion system necessary to return the crew from the Moon requires a figure of merit.

Balancing risks requires people and judgment. There is no unique way to make these types of decisions, but there is a systematic way. While this method cannot assure success, it can ensure against failure. Decision making on an ad hoc and local basis does not consider downstream or across interface impacts. A CRM process will ensure risks are identified and exposed to decisions makers and stakeholders in the Project (especially the Astronaut Office). The process will also ensure transparent rationale is used in making the decision. Figure 2.5-2 illustrates how different assessment methods, shown on the right, supply risk information into risk integration and risk management functions.

A key difficulty in the process is the integration of risks into a coherent framework. Decisions to accept risk should be integrated with the cost and performance dimensions of the constraint "Box," Figure 2.0-2. Decision options must be made using figures of merit (FOMs) that are relevant to the Program requirements, resources, and constraints. A few well-formulated FOMs will simplify the process and aid in transparency. FOMs can be cost, probability of loss of crew, technical performance measures, etc.

Risk is the combination of likelihood and consequence. A key element of risk is uncertainty. Uncertainty can be stochastic in nature, such as the likelihood of completing a mission or can result from a lack of knowledge ("unknown unknowns"). Investments in test programs, simulation modeling, analysis, and engineering model evaluation can reduce this latter uncertainty by increasing knowledge. In this sense, development risks are caused by lack of knowledge and can be retired as the Program progresses towards launch. Identifying and considering uncertainty is an important element in risk assessment. Figure 2.5-3 compares risk assessment methods. As indicated in Section 2.2, the probabilistic methods encourage a deeper understanding of risks sources and what factors are important for reducing likelihood, consequences, and uncertainty.

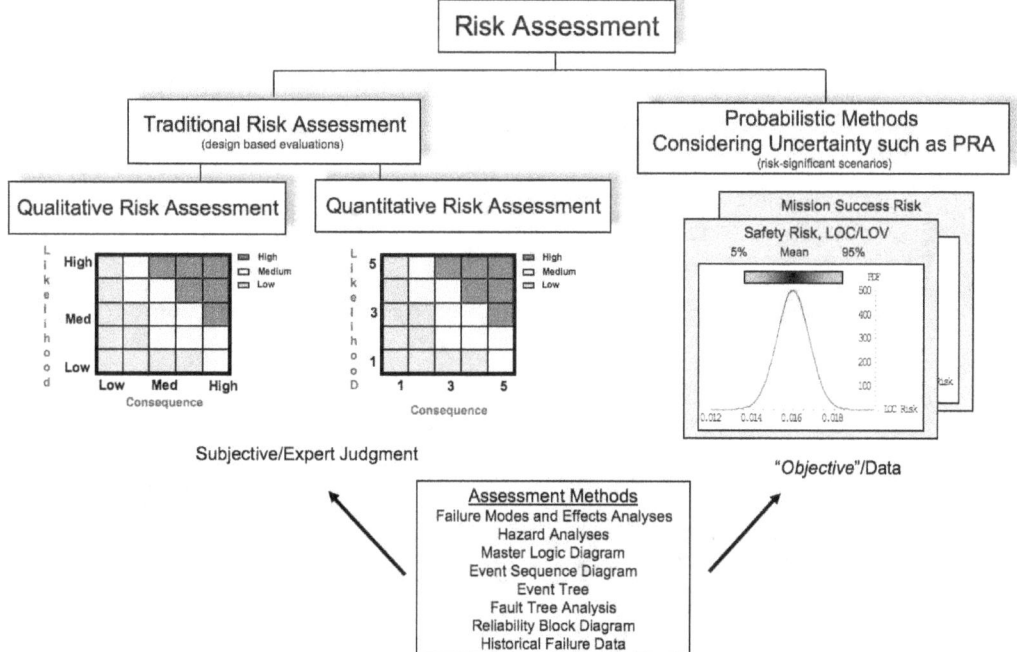

Figure 2.5-3 Qualitative, Quantitative, and Probabilistic Assessment Methods

Adapted from: Schutzenhofer, L.A., Chapter 14.2, "Space Launch and Transportation Risk," in NASA MSFC Course " Space Launch and Transportation Systems: Design and Operations," August 2006

When the launch occurs, the mission risk is comprised of the stochastic uncertainly plus any residual lack of knowledge and unretired mission success risk. Figure 2.5-4 shows a notional example of how total risk, comprised of development risk plus safety/mission success risk, changes over time. This notional example shows how eliminating a test will reduce the development risk (i.e., the test will not fail and affect the schedule by finding and fixing the problem), but increases mission risk. Risk management strategies must take these types of trades into account.

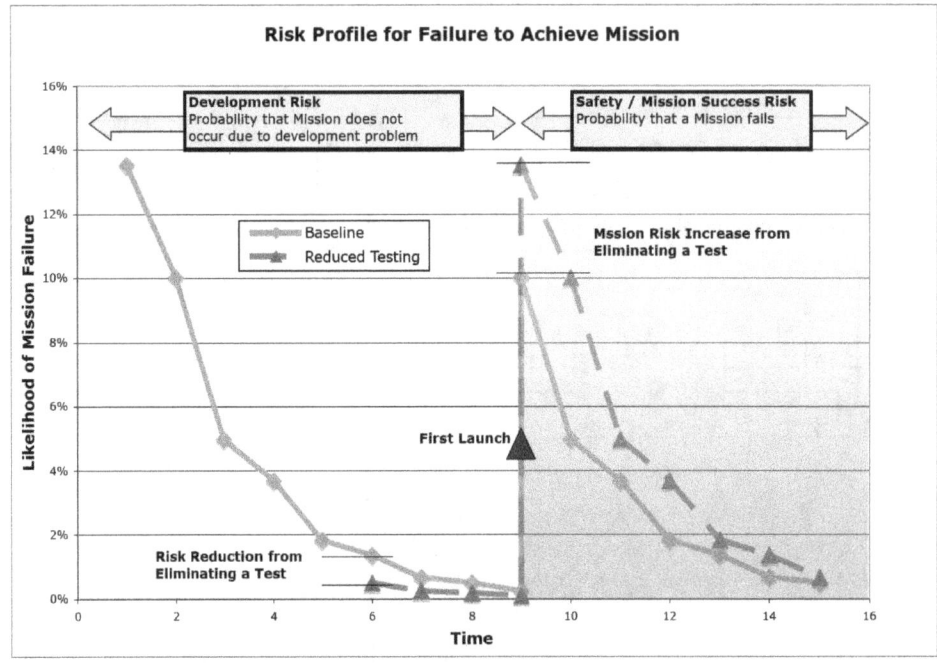

Figure 2.5-4 Life Cycle Risk Profile

2.5.2 Integrating Cost Risk Performance Model into CRM

A top-down cost, risk, and performance model can be used to help managers put lower-level risks in context of the entire program. This model can be used to evaluate the effects of Program elements failing to meet requirements or operate within their constraints. These risk drivers can be used to identify risk criticality used in CRM risk matrices that are developed at lower levels of the Program. Top-down SE must be involved in evaluating the likelihood and consequences of risks that interact across system boundaries. For example, the risk of an inadvertent retro-rocket system fire could be evaluated at the subsystem level due to its obvious consequences, but a failure to fire may depend on the capability of other systems to mitigate the failure and needs to be evaluated to a higher level.

Using a top-down model to evaluate and assess risks overcomes the difficulty in comparing risks that have different likelihood or consequence scales and synergistic effects (interactions between risks are assumed to be independent or constant). For this process to work successfully, risk drivers should be described using a quantitative method so that criteria can be established in the context of the entire system including cost and performance. If the risk becomes too high because of violating cost/schedule, performance, or reliability/safety requirements, the concerns and issues can be elevated.

From a top-down perspective, the risks at lower levels are the likelihood that a requirement or constraint will fail to be met and the impact of the failure on the program. The impact of the failure must be considered using an integrated method, in context with the state of knowledge of the system as a whole. For instance, a reduced performance may not be important if there is sufficient margin in other parts of the system. The performance reduction will then decrease the margin of those systems and increase their risk profile by increasing the consequences if their performance degrades. Therefore, setting risk criteria at lower levels is a dynamic process that must be continuous.

3.0 Safety and Reliability Analysis throughout the Life Cycle

Reliability analyses and risk assessment has been applied in many industries, and the level of acceptance in those industries becomes very high once decision-makers can apply the results of the analysis to their problems. Risk-based design and applications have become an important input to NASA decisions. As confidence in the benefits of the approaches, tools, and techniques is gained, the role of risk assessment becomes central to addressing key safety and reliability issues.

To realize the maximum benefit from reliability analysis, it is essential to integrate the risk and reliability analysts within the design teams. The importance of this cannot be overstated. In many cases, the reliability and risk analysts perform the analysis on the design after it has been formulated. In this case, safety and reliability features are added on or out-sourced rather than designed in. The result is an unrealistic analysis that is not focused on risk drivers, and does not provide value to the design.

To be successful, the techniques described here need to be performed by the design team with the analysts facilitating the process. The models and results need to be owned by the design team. Risk analysts provide a skeptical view that gives a valuable balance to the naturally optimistic view of designers, while their top-down understanding of the role of elements within an integrated system or architecture helps the team focus on risk drivers. This teamwork assures that the rigorous process is followed (reliability practitioner) and that the technical aspects and operational considerations are thoroughly explored (system designer). The designer's involvement also assures that the correct and latest design and operational sequence is analyzed.

Risk and reliability analyses evolve to answer key questions about design trades as the design matures. Reliability analyses lead the growth of information about the system and provide guidance for system risk drivers. Section 2.5 of the NASA Systems Engineering Handbook describes the process of identifying decision drivers, investigating those drivers to make a decision, and then increase resolution for the next level of detail until a full design is realized. Figure 3.1-1 shows key reliability engineering activities and how the focus changes throughout the systems engineering life cycle.

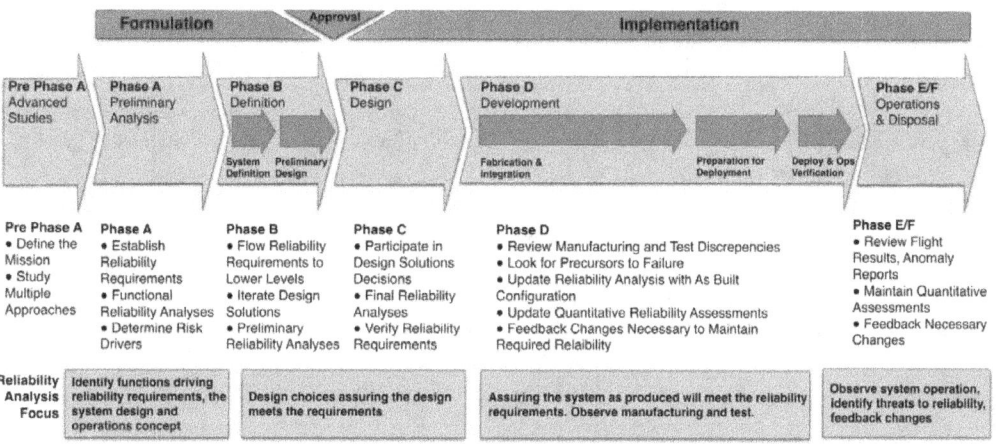

Figure 3.1-1 Systems Engineering Life Cycle with Reliability Focus

The fidelity and level of detail for reliability analyses is driven by the information necessary to understand the risk drivers and create alternatives that will change the cost/performance and risk profile of the system, not by design phasing alone. The key to making this process work is focusing on giving the decision maker information. In this context, information is defined as the "differences that make a difference"[18]. This concept will allow the investigation of many more design options while expending minimal resources necessary to chose the best solution. Keys to successful risk and reliability analysis are:

- The level of detail included in a reliability analysis should be commensurate with the available design information and the level of insight being addressed.

- Too much detail impedes transparency, increases cost, and makes it difficult to manipulate the models. Detail does not equal quality.

- There should be a direct link between heritage and the risk model elements. Credible risk assessments link the design under consideration with empirical evidence from heritage designs.

- The cost-benefit ratio for performing risk assessment is highest during conceptual design.
- The analysis is typically performed in a conservative fashion to identify risk drivers that need to be resolved in ongoing activities. Details developed based on their importance to the risk of the system.

---

[18] Gregory Bateson's definition of information.

- Provide a top-down integrating context for understanding the system.
- Structure of the analysis is driven by the concept of operations, system design, and hazards faced by the system.
- Independent of the artificial boundaries generated by hierarchies of requirements and system decompositions.

Careful peer review of reliability analysis is critical. Review by knowledgeable designers and operators are more important than a review of the math in the calculations themselves. It is important to review analysis inputs, assumptions, and uncertainties, as well as the results. The review needs to resolve any disconnect with the system designers "gut feeling" and what the analysis produces. Outputs should be understandable to design engineering and management. The results should agree with expert intuition or help identify where and why the intuition is flawed.

3.1 Formulation Phases

Reliability analyses are used for risk assessment in the early phases of the design process to develop requirements and design concepts that can meet the requirements. The following aspects of reliability analyses and risk assessments are described for each life cycle phase:

1. Phase Objective
2. Role of Risk and Reliability Analysis
3. Risk and Reliability Techniques
4. Indicators of Proper Reliability and Risk Analysis

Details for each phase will be consolidated into Tables 3.3-1 through 3.3-3 for reference and ease of use.

3.1.1 Pre-Phase A Advanced Studies

Phase Objective

Iterate concepts and requirements to identify feasibility and risk profiles for various options and alternatives. The focus is to identify multiple feasible solutions through the investigation of alternatives, ensuring the overall goals can be achieved within resource constraints.

Role of Risk and Reliability Analysis in Pre-Phase A

The role of risk and reliability in this process is to ensure that risk and reliability objectives are realistic (heritage information, mission complexity, and development risks are properly accounted for), and there is sufficient performance and cost/schedule margin to field a system without compromising reliability. It is the job of risk/reliability analysis/personnel to challenge and document, at least at a top level, inconsistencies in requirements, lifetime, and environments that affect reliability and safety. If these inconsistencies are brought to the forefront then it is easier for the decision makers to understand and accept.

The typical information generated for performance (Make it Work) analysis is insufficient for performing risk assessments at this stage. For a risk assessment to take place, some strawman details need to be in place. These details include mission specific events that potentially drive risks (Launch Vehicle Type, Propulsion Systems employed, and sizing). Critical events (usually involving change in velocity, delta-V, but also involving docking, long-term operations, and operating environments) for the missions must be identified in enough detail to determine how to tailor surrogate elements that represent risk. Another key element in the reliability analysis is the ability of the architecture to compensate for off-nominal conditions and employ mechanisms (diverse or redundant) to save the crew.

With this information, it is possible to gain an understanding of the risk profile of the architecture. This also provides the decision makers with a rough understanding of the driving factors and the likelihood of failures for missions. The analysis in this phase should be conservative in nature such that as analysis matures, the likelihood of being outside the box is low.

Techniques of Risk and Reliability Analysis in Pre-Phase A

Techniques that can be employed for qualitative assessments include Functional FMEA to understand failure modes of mission elements, ESDs to capture operational concepts, scoping risk assessments using surrogate PRA, and heritage data[19,20] to aggregate and estimate risks. If there is uncertainty about the viability of a concept, detailed models may be in order. For instance, the MAST[21] tool is being applied to understand the likelihood of successfully launching two vehicles for a lunar target in a finite amount of time constrained by weather and daylight conditions. This information can be used to set requirements for Lunar Surface Access Module (LSAM) loiter capability/reliability mission definition.

Indicators of Proper Reliability and Risk Analysis in Pre-Phase A

The reliability and risk analysis should be documented in a transparent fashion. The Pre-Phase A column in Tables 3.3-1, 3.3-2, and 3.3-3 summarize some specific activity details necessary for

---

[19] Fragola, J. et al. "Reliability and Crew Safety Assessment for a Solid Rocket Booster/J-2S Launcher", Proceedings of the 2006 RAMS Symposium, Newport Beach, CA.
[20] ESAS Reliability Analysis Chapter 8 and Appendices
[21] The Manifest Assessment Simulation Tool (MAST) is a discrete event simulation environment using Rockwell Software's Arena, ExpertFit by Averill M. Law and Associates and the Microsoft Office suite of Excel, Word, PowerPoint, and Visio. The Manifest Assessment Simulation Tool (MAST) was developed for the Space Shuttle program (Cates 2004; Cates and Mollaghasemi 2005). MAST benefited from the space shuttle model developed in 2001 (Cates et al. 2002).

this life cycle phase. The documentation should include an identification of key assumptions and sensitivities, with uncertainty. Results should identify the reliability drivers and their ranking. The documents should be explicit enough to allow for an independent peer review of the trades and analysis. Just as the analysis was created by the integrated design team, an integrated team of similar composition should perform the peer review. The peer review should focus on reasonableness of the assumptions, modeling and estimates, and identification of risk drivers. The models, data, and documentation should be in a form that can be transmitted to the Phase A team, and referenced and re-used by other studies.

3.1.2 Phase A Preliminary Analysis

Phase Objective

Phase A seeks to converge requirements, design, and operations concepts towards a single feasible solution that meets its performance and risk requirements within acceptable cost and schedule constraints. This is particularly true because this is the very first time the team will commit to a product for a price, so it is important that it is scoped correctly.

Reliability analyses are integral to this effort as described in Section 2.3. An iterative design loop establishes optimal system requirements, top-level architecture, and operations concept. Once a desired mission concept has been identified, then the derived requirements are solidified. The confirmation of a good Phase A is a system architecture meeting its cost/schedule, performance, and risk constraint box.

The elements of the system should have been analyzed to establish confidence that there is sufficient margin to the requirements such that the Phase A design can be built under the constraints and within the requirements. The cost benefit ratio for effective Phase A activities is enormous. For Phase A activities to be effective they need to focus on the drivers.

Role of Risk and Reliability Analysis in Phase A

Risk and reliability analysis help designers understand the interrelationships of requirements, constraints, and resources, and uncover key relationships and drivers so they can be properly considered. The analyst must help designers go beyond the requirements to understand implicit dependencies that emerge as the design concept matures. It is unrealistic to assume that design requirements will correctly capture all risk and reliability issues and "force" a reliable design. System impacts of designs must play a key role in the design. Making designers aware of impacts of their decisions on overall mission reliability is key.

> *Finding Feasible Solutions*
> The search for a feasible solution should begin with the simplest possible version of the mission and vehicles (the de minimis configuration[22]). This version is potentially the

---

[22] De minimis is a Latin expression meaning about minimal things.

most reliable. Once the de minimis mission is identified, it is analyzed to identify risk drivers. The risk drivers are attacked: either eliminated or their impact reduced by providing fault tolerance to reduce their likelihood to an acceptable level. The use of redundancy and diversity is analyzed to provide margin for loss of mission and loss of crew. The mission is then re-analyzed from a performance perspective to make sure it will close. Once the minimal risk mission is defined, then additional performance or return is added to achieve expanded goals. The risk and cost impact of these additions is then calculated so that decision makers can see the effects of adding content to the mission. This process ends when the cost or risk criteria are reached. This is the preferred way to search for a solution since it is very difficult to de-scope missions or reduce capabilities to get in the risk cost box.

Increasing Fidelity of Baseline Models
The baseline models to estimate risk for the Pre-Phase A study are expanded to include more details of the specific options being considered. Human reliability analysis is performed if there are critical human interactions. The additional modeling is focused on risk drivers and the differences between options. To define system requirements, some assumptions must be made. As stated before, the requirements should be flexible, but configuration management is imperative so that analysis is consistent across teams. As more information is gained about specific missions and vehicle configurations, the generic events of the Pre-Phase A models can become more specific. The risk models should be complete in that all risks should be captured even if at a conceptual level.

Risk in Trade Studies
One conceptual difficulty is that in the end, different configurations may well result in equal risks, but require significantly different development and test programs to achieve the same level of reliability. Therefore, safety or mission success risk may not be a differentiator between missions or systems. The difference between elements is reflected in development risk, cost, or performance impacts needed to equalize reliability and safety risks.

It may be impossible to make risks equivalent between concepts no matter how many resources are applied. This makes it impossible to make apples-to-apples comparisons. If the risks cannot be made equal, each system should be made to be cost effective from a risk perspective, and the change in risk noted in the result. This is the reason the trades must be integrated with performance/return and cost. The result of the study is a set of alternatives with different cost risk performance frontiers[23] (surfaces) within the cost risk performance box. Based on the shapes of the frontiers and the margins to the boundary,

---

[23] Finding Pareto Frontiers is particularly useful in engineering. By yielding all of the potentially optimal solutions, a designer can make focused tradeoffs within this constrained set of parameters, rather than needing to consider the full ranges of parameters.

the decision maker can make a choice. If a decision maker does not leave sufficient margin in the choice, system reliability may ultimately suffer due to truncated testing or rushing to launch.

Phase A Inputs

Concept of Operations
- Mission Scenarios
- Contingency Plans

Design
- Alternative design concepts are developed
- Models focused on closing the mission for alternatives Vehicle Configurations and Layouts
- System Block Diagrams

Physics modes for critical aborts

Requirements
- Needs Statement
- Value Function for alternative returns
- Risk Targets
- Time Frame
- Funding Profile
- Constraints
- Requirement Sets need to be derived for alternative missions and vehicles, rather than a given

Techniques of Risk and Reliability Analysis in Phase A

Functional FMEA and preliminary hazard analysis establish failure modes of the hardware. Environmental hazards for the design and failure to complete critical steps of the operations concept are combined with potential failure modes in event sequence diagrams. Physics-based models are combined with the functional FMEA and hazards to understand critical mission failures and potential recovery strategies. In some cases, detailed physics models are needed to understand critical failure modes, their effects, and warning times necessary to establish performance requirements and timing for back-up systems. These details require investigations into accident physics such as those performed by the Simulation Assisted Risk Assessment[24] tool. This tool can be used to establish the likelihood that a crew escape system will perform as required and help set escape requirements. Heritage-based scoping PRA models can be used to evaluate the relative likelihood of scenarios and help identify drivers for mission reliability and

---

[24] Lawrence, S., "Simulation Assisted Risk Assessment," AIAA Paper 2006-0090, Reno, NV, January 2006.

crew safety so that the design can be balanced. Discrete event simulations, Monte Carlo models, and Markov models can be used to evaluate time dependent events than cannot be done analytically.

Indicators of Proper Reliability and Risk Analysis in Phase A

The reliability and risk analysis should be documented in a transparent fashion. The documentation should include an identification of key assumptions and sensitivities, with uncertainty. The Phase A column in Tables 3.3-1, 3.3-2, and 3.3-3 summarize some specific activity details necessary for this life cycle phase.

The documents should be explicit enough to allow for an independent peer review of the trades and analysis. Just as the analysis was created by the integrated design team, an integrated team of similar composition should perform the peer review. The peer review should focus on reasonableness of the assumptions, modeling and estimates, and identification of risk drivers. The peer review should also examine the analysis for completeness to ensure all credible risk drivers are identified. The models, data, and documentation should be in a form that can be transmitted to the Phase B team, and referenced and re-used by other studies.

3.1.3 Phase B Definition

Phase Objective

The objective of Phase B is to define the project in enough detail to establish an initial baseline and preliminary design capable of meeting mission needs and is achievable within technical, cost, schedule, and risk constraints. Sufficient details are developed to formulate and approve a program plan.

Role of Risk and Reliability Analysis in Phase B

The role of risk and reliability in this process is:

- Integrate designs from a reliability point of view.
- Ensure that failure modes are obviated to the maximum extent possible.
- Controls on failure modes are cost effective, and take into account the entire system.
- Ensure that reliability goals can be met.
- Assist in the obviation and mitigation of risks.

Given that the requirements and design concepts were properly defined in Phase A, the purpose of the reliability activities is to demonstrate that the design will have a high likelihood of meeting the requirements. Since information that is more concrete is available about the systems and the concepts of operation, the analysis can begin on terms that are more concrete using established techniques such as those discussed in MIL STD 882[25]. The design and concept of operations should be thoroughly examined for accident initiators and hazards that could lead to mishaps. Conservative estimates of likelihood and consequences of the hazards can be used as a basis for applying design resources to reduce the risk of failures.

Techniques of Risk and Reliability Analysis in Phase B

MIL STD 882 provides excellent guidance on the steps necessary to design a reliable system. Consideration should be given to integrating this process with existing processes. One weakness in the document is the reliance of qualitative measures for assessing the frequency and consequences of mishaps and criteria for acceptable conditions. These deficiencies make it difficult to gain an integrated picture of risks and establish cost benefit of alternative risk mitigation strategies, especially for manned spacecraft systems with their inherent high levels of risk. However, the addition of quantitative measures within the general framework of the guidance provided by the standard significantly improves the ability to assess risks.

Integration with the Design Team

The implementation of this process requires deep knowledge of the element being designed and its context in the overall program. This knowledge cannot be pre-described in requirements. The process depends on an integrated design team that can put the design in context with the entire system and identify as many possible design solutions as possible. Reliability analysts facilitate the process and help focus on risk significant aspects of the design, but are not the sole author of the analysis. The quality of the design is dependent on the quality of the team members and ownership by the engineering organization; it cannot be weighed or measured. The quality of the process must be gauged through direct participation in the design meetings. Insights from design or documentation reviews performed after the design is complete may come too late to efficiently impact program success.

Verification of Compliance with Requirements

Verification that a system is meeting its quantitative risk goals can be accomplished by extending the top down scoping risk assessments to a level of detail consistent with the risks identified in the design. The analysis should be complete, in that all risk drivers have been addressed (design and operational requirements established, and quantified to the extent practical) and risks are acceptable. Results of risk analysis can be used to resolve issues when the application of additional mass or redundancy, necessary to comply with predefined requirements is of questionable or negative value. For a risk analysis to be credible, it must be transparent and connected to the design heritage and physics of the system and its environment. Direct use of handbook data alone for space

---

[25] Department of Defense Standard Proactive for System Safety, Mil STD-882D, 10 February 2000

systems is of questionable value and hurts the credibility of the analysis with engineers and decision makers. It must be recognized that for most of their lives, the reliability of space systems is directly related to flight experience rather than "random" hardware faults. Risk assessments must use a much heritage data as possible and consider the effects of common cause.

Indicators of a Proper Reliability and Risk Analysis in Phase B

In Phase B, the reliability effort shifts from establishing reliability requirements and an architecture to analyzing a preliminary design and assessing its compliance with requirements. The reliability and risk analysis should be documented in a transparent fashion. The analysis should be complete, as quantitatively as practical, in that all risk drivers have been addressed (design/operational requirements established) and risks are acceptable. Results should indicate preliminary design compliance with requirements along with any deficiencies and recommendations for change. The Phase B column in Tables 3.3-1, 3.3-2, and 3.3-3 summarize some specific activity details necessary for this life cycle phase.

Derived requirements generated by any analysis for the final flight design should be included in the requirements management system, along with their rationale. The documentation should include an identification of key assumptions and sensitivities, along with the associated uncertainty. The documents should be explicit enough to allow for an independent peer review of the trades and analysis. The peer review should focus on reasonableness of the assumptions, modeling, estimates, and identification of risk drivers. The peer review should also examine the analysis for completeness to ensure that all credible risk drivers are identified. Just as the analysis was created by the integrated design team, an integrated team of similar composition should perform the peer review. The models, data, and documentation should be in a form that can be transmitted to the Phase C Team, and referenced and re-used by other studies.

3.2 Implementation Phases

The final design, manufacturing, and operation phases use risk assessments and reliability techniques to verify that the design is meeting its risk and reliability goals and help develop mitigation strategies when the goals are not met or discrepancies/failures occur. Human reliability analysis should be performed on critical design, production, assembly, test, and mission operations elements assuring the design is producible with the requisite reliability.

3.2.1 Phase C Design

The purpose of this phase is to establish a complete validated design that is ready to design, code, and manufacture. If the design was correctly defined, this phase results in a design that fully meets the requirements within constraints with margin. Trade studies are to optimize sub-elements of the design focusing on lower level design decisions assuring they remain within their overall requirements and constraints. If the definition phase is defective, SE may need to re-visit trades to establish a new definition that can be realized with acceptable risk.

Just as in the preliminary design process, the analysis matures as the design is completed. If the Phase B analysis and design were performed correctly, the risk drivers and requirements should have already been identified along with effective solutions. This phase of the analysis continues to validate the system design from a risk perspective by focusing on the final design details ensuring that all risk drivers have been identified and design details do not degrade or introduce new risk drivers.

The detailed application of risk analysis tools (Fault Trees, FMEAs, and Risk Assessments) are used to document the compliance of the design with the requirements and ensure that the risk drivers for the system continue to be identified and resolved. Process FMEAs are used to assure that manufacturing and integration processes are sufficiently robust. The documentation and models serve as a way to capture the design basis and rational for use in resolving problems and implementing design changes during development and operation.

Indicators of a Proper Reliability and Risk Analysis in Phase C

Reliability analyses should be complete and show that reliability requirements are met. Reliability analysis of the final design should be documented in a fashion that allows independent peer review. Results should indicate compliance with requirements along with any deficiencies and recommendations for change. The Phase C column in Tables 3.3-1, 3.3-2, and 3.3-3 summarize some specific activity details necessary for this life cycle phase.

The various reliability analyses used to verify requirements should be consistent with each other. The quantitative analyses should support the qualitative analyses and compliance with fault tolerance assessments. For example, the qualitative analyses should support design for minimum risk rationale.

Analysis results should include an identification of key assumptions and sensitivities with uncertainty. The documents should be explicit enough to allow for an independent peer review of the design choices and analysis. The peer review should focus on reasonableness of the assumptions, modeling, estimates, and identification of risk drivers. The peer review should also examine the analysis for completeness to ensure that all credible risk drivers are identified. Just as the analysis was created by the integrated design team, an integrated team of similar composition should perform the peer review. The models, data, and documentation should be in a form that can be used later in the life cycle and referenced and re-used by other studies.

3.2.2 Phase D Development

The purpose of this phase is to build the subsystems and integrate them into a complete system, developing confidence that the system will be able to meet its requirements. The system should then be deployed, ensuring that it is ready for operations.

The role of reliability in this phase is to monitor the test and integration program, identify precursors to failure, and help resolve problems as they are identified. This will assist in discovering the failure root causes, suggested corrective actions, and verifying that these actions have been implemented properly and have been effective. The nature of space programs makes it difficult to use statistics from tests to develop estimates of system reliability. These statistics must be gleaned from operational data and subjectively extended to the system in question.

MIL STD 882 provides the following guidance for testing a safe design:

> "A.4.4.6.1 Testing for a safe design. Tests and demonstrations must be defined to validate selected safety features of the system. Test or demonstrate safety critical equipment and procedures to determine the mishap severity or to establish the margin of safety of the design. Consider induced or simulated failures to demonstrate the failure mode and acceptability of safety critical equipment. When it cannot be analytically determined whether the corrective action taken will adequately control a hazard, conduct safety tests to evaluate the effectiveness of the controls. Where costs for safety testing would be prohibitive, safety characteristics or procedures may be verified by engineering analyses, analogy, laboratory test, functional mockups, or subscale/model simulation. Integrate testing of safety systems into appropriate system test and demonstration plans to the maximum extent possible."

Reliability and risk analysis and techniques can be used to advise systems engineers on the efficacy of test protocols for achieving reliability goals. The risk and reliability analysis should be part of the problem resolution process to ensure that analysis and assumptions developed in early phases are correct, and can help ensure that problem resolutions are effective and do not degrade overall system reliability.

Indicators of a Proper Reliability and Risk Analysis in Phase D

In Phase D, the reliability efforts shift from verifying the final design to following the manufacturing and test efforts assuring that construction and test techniques are consistent with the intended reliability. The Phase D column in Tables 3.3-1, 3.3-2, and 3.3-3 summarize some specific activity details necessary for this life cycle phase.

Manufacturing and test results may indicate that updates to reliability models and assumptions are required. Members of the reliability team should be involved in discrepancy closure on hardware identified as mission critical.

Changes to the final design may also be necessary and should be reviewed by the reliability team.

3.2.3 Phase E Operation

The purpose of this phase is to meet the initially identified need. The products of this phase are the operational capability achieved, and science and experiences returned from the mission.

The role of reliability in this phase is to monitor problem reports, and identify risk significant issues and their possible resolution. Precursor analysis can be used to verify reliability estimates used to define the program.

Shortcomings in existing data collection systems have hampered the ability to develop quantitative estimates of reliability parameters and significantly increased uncertainty and forced analysis to be based on handbook data. The inability to develop reliability insights from the historical data illustrates the need for significant improvement in data collection and analysis. Limited experience with space systems makes this data even more precious. *NASA should ensure that issues of usability and data quality are addressed in new versions of PRACA and that end user needs and requirements are taken into consideration.*

Indicators of a Proper Reliability and Risk Analysis in Phase E

In Phase E, the reliability efforts shift from following the manufacturing and test processes to assuring the system is operated in a manner consistent with its intended and assumed operations plan. The Phase E column in Tables 3.3-1, 3.3-2, and 3.3-3 summarize some specific activity details necessary for this life cycle phase.

Flight results may indicate that updates to reliability models and assumptions maybe in order. Members of the reliability team should be involved in flight anomaly and discrepancy closure.

Changes to the system or its operations plan may also be necessary and should be reviewed by the reliability team.

3.3 Application of Risk Analysis Tools and Techniques

The focus and application of reliability analyses tools and techniques varies throughout the system life cycle as life cycle phase objectives change, information about the system matures, and issues occur. Tables 3.3-1, 3.3-2, and 3.3-3 describe[26] approaches for reliability and risk analysis. The descriptions are organized by mission phase describing the scope of effort and types of analyses as they evolve and the program progresses along the life cycle. The tables are organized into three major sections across the life cycle phases.

- Reliability Assessment activities have been organized into the following five areas and described in Table 3.3-1.
    1. **Requirements** are evaluated, defined, and flow into lower levels of the system hierarchy.
    2. **Validation and verification** determines whether or not the requirements for a program reflect the program intentions and whether or not the developing design is in compliance with the requirements. Validation and verification can be assessed qualitatively or quantitatively.

---

[26] based on a taxonomy and description originally developed for the NASA Engineering for Complex Systems project at Ames Research Center.

3. **Evaluation of alternatives** refers to trade studies, wherein alternative design solutions are evaluated to determine their potential risk, performance, and cost. The intent of these studies is to identify design features that discriminate among these parameters and regimes that appear to encompass alternative design alternatives.

4. **Investigations** are forensic applications that identify likely causes for problems. It is important for the team to utilize a "physics of failure" approach when analyzing critical failures in addition to using the "scientific method" when faced with unknown or uncertain failure mechanisms.

5. **Risk integration** provides management with a view of how risks are combined across the program and helps to focus resources on drivers.

6. **Planning and resource allocation** provides information to project management on where project resources can be best applied to reduce risk. The objective to adjust and tailor the reliability and risk analyses as needed to identify risks and reduce the most significant ones.

- Analyses Techniques and Methods; see Table 3.3-2.
    1. **Event Sequence Diagrams/Event Trees** are models that describe the sequence of events and responses to off nominal conditions that occur during a mission.
    2. **FMEAs** are a bottom up analysis that identify the types of failures that can occur within a system and identify the causes, effects, and mitigating strategies that can be employed to control the effects of the failure.
    3. **Qualitative Top-Down Logic Models** identify how failures within a system can combine to cause an undesired event.
    4. **Quantitative Logic Models (Probabilistic Risk Assessment)** extend the qualitative models to include the likelihood of failure. This typically involves developing failure criteria based on system physics, and employing statistical techniques to estimate the likelihood of failure along with uncertainty.
    5. **Reliability Block Diagrams** are diagrams of the elements of a system are combined to provide a function.
    6. **Preliminary Hazard Analysis (PHA)** is performed early based on the functions performed during the mission. PHA is a "What if" process that considers the potential hazard, initiating event, effect, and potential corrective measures and controls. The objective is to determine if the hazard can be eliminated, and if not, how it can be controlled.
    7. **Hazard Analysis (HA)** evaluates the completed design. HA is a "What if" process that considers the potential hazard, initiating event, effect, and potential

corrective measures and controls. The objective is to determine if the hazard can be eliminated, and if not, how it can be controlled.

8. **Human Reliability Analysis** is a method to understand how human failures can lead to system failure and estimate the likelihood of those failures.
9. **Probabilistic Structural Analysis** provides a way to combine uncertainties in materials and loads can combine to cause failure of a structural element.
10. **Sparing/Logistics Models** provide a means to estimate the interactions of systems in time. These models include ground-processing simulations and mission campaign simulations.

- Information for risk and reliability analysis must be available with the appropriate level of detail and maturity at the time of the analysis. See Table 3.3-3.
    1. **Top Level Requirements** representing needs, objectives, and constraints describe the overall objective of the program requirements and budgetary and time constraints. These begin at a high level and are traded during Phase A and verified during later phases.
    2. **Operations Concept** describes the sequence of operations for the mission. Mission design can fundamentally alter the risk and reliability of a program. The manner in which architecture and design matures for a new system is a key driver.
    3. **Architecture and Design** affects risk through the fundamental reliability of its elements and the ability to employ of diverse capabilities for abort. Employing new technology impacts development risk and reliability during missions.
    4. **Requirements** are developed and documented to fix a design and its assumptions so that design and analysis can proceed and remain coherent. These are captured as derived requirements representing design decisions critical to lower levels of the system hierarchy.

Table 3.3-1 Reliability Assessment Functions Along the Systems Life Cycle

	Formulation				Implementation	
	Advanced Studies (Pre-A)	Preliminary Analysis (A)	System Definition (B)	Preliminary Design (B)	Final Design (C)	Development (D)	Mission Operation (E)
Systems Life cycle Phase Objective	Evaluate Multiple Solutions	Choose a Single Best Solution		Demonstrate via a preliminary design that the system is "In the Box". Show the solution works, is safe, and is affordable	Demonstrate the Final Design will work as intended and is producible	Make it and Test it	Fly It
Reliability Team Focus	• Participating in SE Trades, select goals • Define Reliability Requirements • Identify concepts & alternatives • "Functional" Reliability PRA Analysis and ranking of alternatives with detail where necessary • Define additional study • Tailor Reliability Approach to Program needs	• Collect relevant historical data • Analyze solution Safety, Reliability, Risk based on Functions • Assess new technology • Flow Reliability Requirements to lower levels • Iterate solution based on feedback • Identify Risk Drivers • Draft Reliability Program Plan, inputs to Systems Engineering, Safety and Performance Assurance Plans	• Flow Reliability Requirements to lowest levels • Participate in design solution trades • Perform preliminary analysis of the design showing it meets requirements • Preliminary design analyses (Worst Case, Parts Stress, Load Margin) necessary for reliability • Preliminary FMEA, CIL, Hazard Reports		• Participate in final design choices • Perform final analysis of the design showing it meets requirements • Final design analyses (Worst Case, Parts Stress, Load Margin) necessary for reliability • Final FMEA, CIL, Hazard Reports	• Review manufacturing discrepancies tracking/trending and MRBs • Review Test anomaly reports • Look for precursors to failure • Maintain and update quantitative reliability risk assessments • Feedback changes necessary to maintain required reliability	• Review flight results, tracking/trending, anomaly reports • Look for precursors to failure • Maintain and update quantitative reliability risk assessments • Feedback changes necessary to maintain required reliability

NASA Engineering and Safety Center Technical Report

Document #: RP-06-108
Version: 1.0

Design Development Test and Evaluation (DDT&E) Considerations for Safe and Reliable Human Rated Spacecraft Systems

	Formulation			Implementation		
	Advanced Studies (Pre-A)	Preliminary Analysis (A)	System Definition (B)	Preliminary Design (B)	Final Design (C)	Development (D)	Mission Operation (E)
Requirements	Participate in definition of acceptable risk. Establish top level reliability requirements	Help define Mission Level reliability requirement 1) Fault Tolerance, 2) Quantitative, 3) Processes. Requirements Flow Down	Participate in defining reliability and QA Requirements, manufacturing tolerances, and processes critical for system reliability. Flowdown of reliability requirements		Identify and track any changes	Identify and track any changes. Follow manufacturing and testing to assure required processes are followed	Follow operations to assure required processes are followed
Validation and Verification	Are the reliability goals reasonable for the mission?	Does the concept meet the goals?	Does the preliminary design meet the reliability requirements?		Verify qualitative goals (design criteria). Verify final design meets reliability requirements	Verify hardware is produced with the quality necessary for required reliability. Track problem reporting for precursors and warnings that reliability may be compromised	Are the goals being met?
Reliability Analyses	Tailor necessary reliability analysis to program unique needs. Perform high level analyses at functional level	Choose appropriate analyses techniques. Perform high level analyses at functional level with design detail as necessary. Feedback of risk information	Choose appropriate analyses techniques. Refine modeling data and its uncertainty. Perform analyses based on preliminary design. Feedback of risk information		Choose appropriate analyses techniques. Perform analyses of final design. Feedback of risk information	Update analysis per changes and test results. Update models and uncertainty as necessary based on test failures / results	Update analysis per flight results. Update models and uncertainty as necessary based on flight failures / results
Evaluation of Alternatives	Help define goals and evaluation criteria. Establish feasible alternatives	Participate in trade to select a single best solution	Trade studies leading to design choices, and optimized systems. Interface selection. Component selection		Design details, interface details. Fabrication process alternatives	Discrepancy and problem resolution alternatives	Discrepancy and problem resolution alternatives
Investigation (Physics of Failure and use of Scientific Method)					What are proximate and root causes of development unit test failures	Proximate and root cause of flight unit failures	Proximate and root cause of flight failures

	Formulation			Implementation		
	Advanced Studies (Pre-A)	Preliminary Analysis (A)	System Definition (B)	Preliminary Design (B)	Final Design (C)	Development (D)	Mission Operation (E)
Risk Integration *(Integrating Reliability Analysis Outputs with Risk Assessments)*	Identify Development Challenges, and system goals that are at or beyond the state of the art	Identify the risk contributing mission elements, and development risks	Evaluate risk drivers (Mission, Development)		Identify relevancy of design, Identify test and verification processes to address risk drivers	Evaluate test results for indications that design and environment are not fully understood	Evaluate operational experience to test operational assumptions and risk estimates
Planning and Resource Allocation	Define reliability analysis and resource needs based on mission objectives	Plan and tailor reliability activities based on mission needs and requirements. Reliability Program Plan	Update Reliability Plan as necessary		Update Reliability Plan as necessary	Update Reliability Plan as necessary	Update Reliability Plan as necessary

Table 3.3-2 Reliability and Risk Assessment Tools and Techniques

	Formulation				Implementation		
	Advanced Studies (Pre-A)	Preliminary Analysis (A)	System Definition (B)	Preliminary Design (B)	Final Design (C)	Development (D)	Mission Operation (E)
Event Sequence Diagrams/ Event Trees	Identify main risk drivers in the mission	Details added mission events that drive mission risk	Details included to model specific system implementations	Fully Developed based on preliminary design	Updated with final design information and operations plans	Updated with design changes and final operations procedures	Update with changes to flight system and operations plan / procedures
FMEA / CIL	High functional level to identify risk drivers	Functional/ System Level depending on risk importance	Preliminary design Design unknowns based on functions or similar units	Final design Waivers for non compliances	Component Level	Update with changes to flight system and operations plan	Update with changes to flight system and operations plan / procedures
Qualitative Top Down Logic Models *(Master Logic Diagram, non Quantified Fault Tree Analysis)*	Highest Functional/ System level depending on risk importance	Functional/ System level depending on risk importance	System and subsystem level depending on risk importance	System/Component Preliminary Design depending on availability of design information	System/Component based on Final Design	Developed to cause level for causal analysis	Developed to cause level for causal analysis
Quantitative Logic Models *(Probabilistic Risk Assessment PRA, Monte Carlo Simulations)*	Highest level assessment supporting concept and definition of acceptable risk	Level of detail needed to support Event Trees and Event Sequence Diagrams Assessment of candidate approach Examine Historical Data for failure rates and uncertainty	Functional/ System Level depending on risk importance Identify risk drivers	Based on preliminary design to assess compliance to requirements Identify risk drivers	Fully Developed Refine failure likelihood data for failure rates and uncertainty	Fully Developed, Causal Event Trees for failures Update models based on observed failure rates	Fully Developed, Causal Event Trees for failures, Automated calculation for risk monitoring, logistics studied Update models based on observed failure rates

NASA Engineering and Safety Center Technical Report

Document #: RP-06-108
Version: 1.0

Design Development Test and Evaluation (DDT&E) Considerations for Safe and Reliable Human Rated Spacecraft Systems

	Formulation				Implementation	
	Advanced Studies (Pre-A)	Preliminary Analysis (A)	System Definition (B)	Preliminary Design (B)	Final Design (C)	Development (D)	Mission Operation (E)
Reliability Block Diagrams		Level of detail needed to support Event Trees	Functional/ System Level depending on risk importance	System/Component Level based on preliminary design	Fully Developed to final design	Fully Developed, Causal Level Trees for failures	Fully Developed, Causal Level Trees for failures
Preliminary Hazard Analysis (PHA)		Preliminary Hazard Analysis based on functions by Safety team with subsystem and reliability participation		Preliminary Hazard Analysis by Safety team based on Preliminary Design with subsystem and reliability participation. Preliminary Hazard Reports			
Hazard Analysis (HA)					Hazard Analysis by Safety team based on Final Design with subsystem and reliability participation. Generation of Hazard Reports defining Hazard Controls	Monitor and track Hazard Controls captured in Hazard Reports	Monitor and track Hazard Controls captured in Hazard Reports
Human Reliability Analysis	Limiting Key Mission Activities	Key Mission Activities/Events	Key Recovery Actions, Manufacturing Assembly Processes	Analysis of process controls for Design to Minimum Risk Systems	Analysis of process controls for systems, events without diverse back-up	Process FMEA for critical elements, and mission events for human interactions	Analysis of precursor events for human interactions
Probabilistic Structural Analysis	Key Mission activities or requirements	Analysis of concepts for damage tolerance	Functional/ System Level depending on risk importance	System/Component Level depending on risk importance availability of design information	Fully Developed	Developed to cause level for causal analysis	Developed to cause level for causal analysis
Sparing / Logistics Models	Simple	Level of detail necessary to evaluate the infrastructure necessary to support a concept		Models supporting the preliminary design	Detailed Models supporting the system design	Update based on test history	Detailed models Updated with flight history

Table 3.3-3 Information Required for Reliability Assessments

	Formulation				Implementation	
	Advanced Studies (Pre-A)	Preliminary Analysis (A)	System Definition (B)	Preliminary Design (B)	Final Design (C)	Development (D)	Mission Operation (E)
Top Level Requirements (Needs, Objectives, and Constraints)	Basic objectives of the mission, time frame, rough budget impact Understanding and formulation of Acceptable Risk	Baselined mission requirements, additional requirements that add value by driving the system solution in a desired direction	Track Changes	Track Changes	Track Changes	Track Changes	
Operations Concept	Number of missions, and mission description including propulsion events, and time elapsed	Mission descriptions that include propulsion events, landing, docking, EVA and abort modes	Refined level of detail sufficient to evaluate preliminary design	Final Operations "Concept" becomes Draft Operations "Plan" Sufficient detail to verify final design	Final Operations Plan	Track Changes, look for reliability impacts	Track Changes, look for reliability impacts
Architecture & Design	Alternative design concepts that will meet mission objectives, and heritage reliability information, technical challenges	Functional Block diagram High level design block diagram Hardware software allocation of function Level of redundancy, engine out, abort capability	Final design block diagram Software functional design	Preliminary design, parts lists	Final design drawings and parts lists Software preliminary code	Track Changes, look for reliability impacts Final Software Code	

	Formulation			Implementation		
	Advanced Studies (Pre-A)	Preliminary Analysis (A)	System Definition (B)	Preliminary Design (B)	Final Design (C)	Development (D)	Mission Operation (E)
Requirements	High Level enough to define a successful program and where necessary to drive the system design	Reliability Requirement 1) Fault Tolerance, 2) Quantitative, 3) Processes Requirements Flowdown Complete to System level and subsystem level where necessary	Requirements Flowdown Complete to subsystem level	Requirements Flowdown Complete to components	Track Changes, look for reliability impacts	Track Changes, look for reliability impacts	

3.4 Key Issues in Quantifying Risk

Historical data shows that system reliability is driven by generic (potentially common cause) failures, and that these failures causes generally decrease (reliability growth) as the system matures. The low flight rates of space systems make it difficult to achieve a steady state reliability that results from random parts failures alone. Therefore, quantitative analysis should take these factors into consideration if risk is to be accurately characterized. This section discusses common cause failures, system maturity, and the use of heritage data to calibrate models.

3.4.1 Reliability Analysis Consideration of Common Cause Failures

Reliability models of systems sometimes make the assumption that all failures are statistically independent; that is, the probability of a failure is the same regardless of the state of the system. This assumption typically results in incredibly high reliability estimates of systems that have multiple strings of identical redundant components. Experience shows that this assumption is usually not valid, and a conditional probability (common cause) should be used to assess the probability of a second failure. Failure history described in Section 1.2 indicates a significant fraction of flight failure causes are "generic" in nature that can result in the failure of multiple units. These kinds of "generic" causes are broadly identified as "common cause" in the following sections.

To thoroughly assess system reliability, there are two "knobs" on the reliability models: the "statistically independent" or random failure and "dependent" or "common cause" failures. Using both terms helps to evaluate the effectiveness of failure mitigation strategies and avoids an overdependence on redundancy. Mitigation of common cause failure usually requires a diverse path. The benefit of the diverse path is directly correlated with its independence from the failure cause.

Figure 3.4-1 shows probability estimates for conditional probabilities for failures of second through fourth strings of equipment. The lower left corner shows a notional probability that a first string will fail. The probability of a first string failure is seen to be relatively small. The three broad bands (one green and two grey) above and to the right of the first string failure probability represent a set of estimates for the probability of a second string failing given that the first string has failed; the probability of a third string failing given that two strings have failed; and the probability of a fourth string failing given that three strings have failed, respectively.

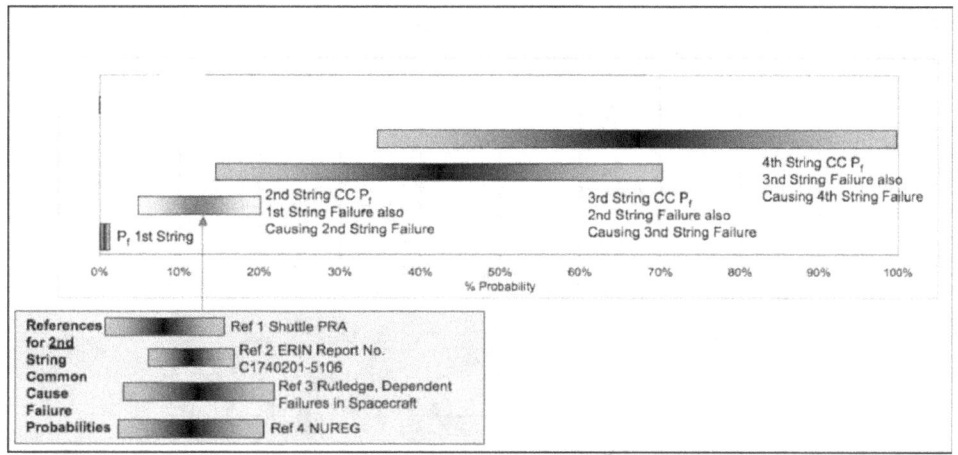

Figure 3.4-1 Conditional Probabilities of Multiple Failures of Identical Components

Some studies have been performed on second string failures (ref. 2 and ref. 3) on Space Shuttle components and other non-aerospace sources (ref. 3). As shown in the green block in Figure 3.4-1, a range of failure probabilities was determined in each of these references that include uncertainty. The four references are all in relatively good agreement to allow an estimating range of probabilities for the second string failures shown as the green line in Figure 3.4-1.

Estimates for the third and fourth string failures for aerospace components are more difficult to find. No studies have been conducted to provide definitive estimates for third and fourth string failures for spacecraft components since such a high degree of redundancy is rare. However, extensive studies have been performed by the commercial nuclear power industry in this area that provides some insight to allow reasonable estimates to be made. Although the industries requirements are very different, both require very high levels of safety and, therefore, both use stringent manufacturing, inspection, and testing processes to ensure high reliability. Data derived from the nuclear industry (NUREG-5485) for the third and fourth failures result in estimates spanning 0.57 for the third string and 0.65 for the fourth string. Using these generic nuclear industry estimates, engineering judgment was applied to establish credible ranges for these parameters as shown in Figure 3.4-1. Note the broad bands signifying the large uncertainty in these estimates.

A key insight from this analysis is that a diverse system (even if it is only 90 percent effective), will provide significantly more benefit than third or fourth set of identical channels

Figure 3.4-2 provides a qualitative example showing how common cause effects might affect the estimates of system failure probability. The probability of failure for a single string system with a failure rate of 20 failures per million hours over a 5000-hour mission is set at a reference of one. If a second string is added, the failure rate is reduced by a factor of 10, reliability improved by a factor of 10, red bar. Considering common cause, the improvement would be limited to five, blue bar. For a third string, the reliability improvement would be a factor of 100 considering just random failures and limited to 9 when considering common cause. For the example shown in Figure 3.6-2, additional redundancy beyond three strings has a limited value when considering common cause effects. Considering just random failures, reliability improvements continue with additional redundancy.

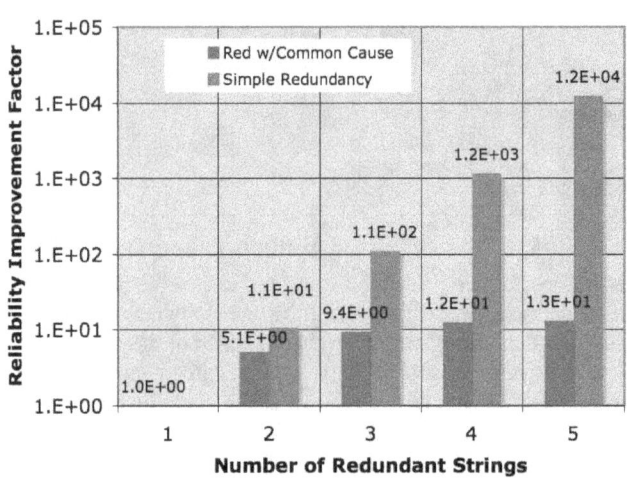

Figure 3.4-2 Effects of Redundancy and Dependent Failures on Reliability

3.4.2 Maturity Modeling

Understanding reliability growth is essential to account for the infant mortality and infant mortality problems with systems. The high cost of space flight makes it difficult to achieve the reliability typically predicted by component-based models. This effect must be taken into account since missions will often occur before maturity is achieved in essential systems. Recent studies such as ESAS[27] evaluate the transition from initial system tests to operation of a mature system. These studies take into account the maturity and complexity of systems and attempt to model the transition of the risk from design and manufacturing problems occurring early in the program to steady state or plateau reliability reached after these problems are sorted out. Figure 3.4-3 shows analytical estimates of reliability of US Launch Vehicles improving over time based on flight experience.[28]

---

[27] NASA's Exploration Systems Architectural Study, Final Report, TM-2005-214062 November 2005, Chapter Section 8.5 Architecture Model.
[28] Space Launch Vehicle Reliability, I-Shih Chang, Aerospace Crosslink, Winter 2001.

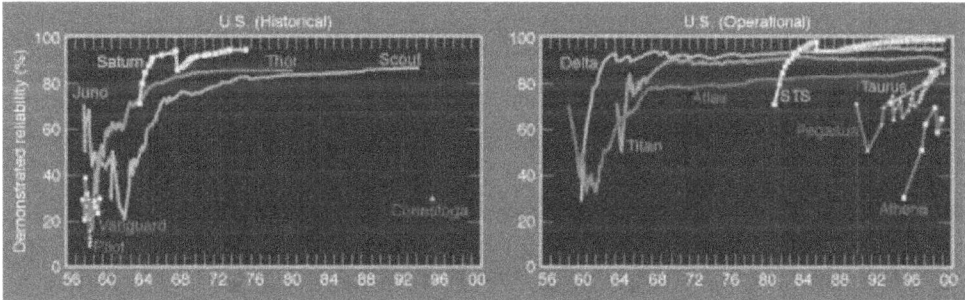

Figure 3.4-3 Demonstrated Launch Vehicle Reliability Improvement with Maturity
Space Launch Vehicle Reliability, I-Shih Chang, Aerospace Crosslink, Winter 2001

3.4.3 Heritage Data

It is important for reliability analysis to be compared to historical data (demonstrated reliability) for similar systems. This comparison can help calibrate the analysis and provides additional confidence that the analysis is not too optimistic. It is good for the demonstrated reliability analysis to describe the individual failure events that go into the calculation. The analysis should understand the relevance of the failure to the system and provide rationale for and discounts used to reduce the impact of the events. This analysis should also account for the potential sparseness of the data set by including an estimate of the uncertainty associated with the estimates.

Acronyms

CAIB	Columbia Accident Investigation Board
CIL	Critical Items List
COTS	Commercial Off-the-Shelf
CRM	Continuous Risk Management
DDT&E	Design, Development, Test, and Evaluation
DoD	Department of Defense
EEE	Electrical, Electronic, and Electromechanical
EPR	Engineering Peer Review
ESAS	Exploration Systems Architecture Study
ESD	Event Sequence Diagram
FMEA	Failure Mode and Effects Analysis
FOM	Figure of Merit
GSE	Ground Support Equipment
HA	Hazard Analysis
ICD	Interface Control Document
ISS	International Space Station
JSC	Johnson Space Station
MCM	Multi Chip Modules
MRB	Material Review Board
MSFC	Marshall Space Flight Center
NESC	NASA Engineering and Safety Center
PBS	Product Breakdown Structure
PHA	Process Hazards Analysis
PLOA	Probability of Loss of Aircraft
PRA	Probabilistic Risk Assessment
RFA	Request for Action
RID	Review Item Discrepancy
S&MA	Safety and Mission Assurance
SE	Systems Engineering
SEO	Systems Engineering Office
SFA	Space Flight Awareness
SPRT	Super Problem Resolution Team
SSP	Space Station Program
STS	Space Transportation System
TRL	Technology Readiness Level
WBS	Work Breakdown Structure

Appendix A. NESC Request Form

NASA Engineering and Safety Center Request Form		
Submit this ITA/I Request, with associated artifacts attached, to: **nrbexecsec@nasa.gov**, or to NRB Executive Secretary, M/S 105, NASA Langley Research Center, Hampton, VA 23681		
Section 1: NESC Review Board (NRB) Executive Secretary Record of Receipt		
Received (mm/dd/yyyy h:mm am/pm) 10/7/2005 12:00 AM	Status: New	Reference #: 05-173-E
Initiator Name: Michael Bloomfield	E-mail: michael.j.bloomfield@nasa.gov	Center: JSC
Phone: ( )- - , Ext	Mail Stop:	
Short Title: Reliability and Redundancy Trades		
Description: See attached email chain		
Source (e.g. email, phone call, posted on web): email		
Type of Request: Assessment		
Proposed Need Date:		
Date forwarded to Systems Engineering Office (SEO): (mm/dd/yyyy h:mm am/pm)		
Section 2: Systems Engineering Office Screening		
Section 2.1 Potential ITA/I Identification		
Received by SEO: (mm/dd/yyyy h:mm am/pm): 10/7/2005 12:00 AM		
Potential ITA/I candidate? ☒Yes ☐No		
Assigned Initial Evaluator (IE): Julie Kramer White		
Date assigned (mm/dd/yyyy): 10/14/2005		
Due date for ITA/I Screening (mm/dd/yyyy):		
Section 2.2 Non-ITA/I Action		
Requires additional NESC action (non-ITA/I)? ☐Yes ☐No		
If yes:		
Description of action:		
Actionee:		
Is follow-up required? ☐Yes ☐No If yes: Due Date:		
Follow-up status/date:		
If no:		
NESC Director Concurrence (signature):		
Request closure date: .		
Section 3: Initial Evaluation		
Received by IE: (mm/dd/yyyy h:mm am/pm):		
Screening complete date:		
Valid ITA/I candidate? ☐Yes ☐No		
Initial Evaluation Report #: NESC-PN-		
Target NRB Review Date:		

NESC Request Form   Page 1 of 4
NESC-PR-003-FM-01, v1.0

NESC Request Number: 05-173-E

Section 4: NRB Review and Disposition of NCE Response Report			
ITA/I Approved: ☐Yes ☐No	Date Approved:		Priority: - Select -
ITA/I Lead: , Phone ( ) - , x			
Section 5: ITA/I Lead Planning, Conduct, and Reporting			
Plan Development Start Date:			
ITA/I Plan # NESC-PL-			
Plan Approval Date:			
ITA/I Start Date	Planned:	Actual:	
ITA/I Completed Date:			
ITA/I Final Report #: NESC-PN-			
ITA/I Briefing Package #: NESC-PN-			
Follow-up Required? ☐Yes ☐No			
Section 6: Follow-up			
Date Findings Briefed to Customer:			
Follow-up Accepted: ☐Yes ☐No			
Follow-up Completed Date:			
Follow-up Report #: NESC-RP-			
Section 7: Disposition and Notification			
Notification type: - Select -		Details:	
Date of Notification:			
Final Disposition: - Select -			
Rationale for Disposition:			
Close Out Review Date:			

```
X-Sender: r.r.roe@pop.larc.nasa.gov (Unverified)
>X-Mailer: QUALCOMM Windows Eudora Version 6.1.1.1
>Date: Fri, 07 Oct 2005 13:27:27 -0400
>To: julie.a.kramer@nasa.gov,
>   d.m.schaible@larc.nasa.gov,
>   k.d.cameron@larc.nasa.gov
>From: "Ralph R. Roe" <Ralph.R.Roe@nasa.gov>
>Subject: Fwd: Reliability
>
>Julie, Dawn, Ken,
>
>Please read Bloomer's not below. I think this would be a very good study
>for us to perform independent of the new program. Julie since you did so
>much work on Human Rating I would like you to lead a small team with help
>from System Engineering, S&MA (reliability experts) and our Discipline
>Experts. Dawn and Ken please help Julie with the appropriate experts from
>your areas. Lets put together a plan on how to attack this, have some
>discussions with Bloomer and the CB folks and we will propose it to Doc
>Horowitz. Thanks
>
>Ralph
>
>
>
>>From: "BLOOMFIELD, MICHAEL J. (JSC-CB) (NASA)"
>><michael.j.bloomfield@nasa.gov>
>>To: Ralph.R.Roe@nasa.gov
>>Subject: Reliability
>>Date: Fri, 7 Oct 2005 11:02:16 -0500
>>X-Mailer: Internet Mail Service (5.5.2657.72)
>>X-Proofpoint-Spam-Details: rule=notspam policy= score=0 mlx=0
>>adultscore=0 adjust=0 engine=2.5.0-05091301 definitions=2.5.0-05100700
>>
>>Ralph,
```

>>   There have been some heated discussions in our office (CB) about
>> the best way to make sure that the next vehicle is highly reliable. I
>> could forward you the e-mails for your reading entertainment, but I'm
>> not sure your in-box is sufficiently large! :-)
>>
>>   In a nutshell, the discussions always center around the trade
>> between levels of redundancy and the cost (weight, $$, complexity)
>> associated with redundancy in order to achieve a highly reliable
>> vehicle. I was on the team that put together the latest version of the
>> HRR, and the HRR requires two fault tolerance to loss of life, unless
>> you can show that the redundancy decreases reliability or the system is
>> highly reliable without redundancy.
>>
>>   As far as I know, I don't think anyone at NASA has decided which
>> type of systems are highly reliable without redundancy and which systems
>> require redundancy in order to achieve high reliability. I think most
>> would agree, however, that there are different standards for different
>> systems. For example, I think most folks will agree that pressure
>> vessels built to a safety factor of 1.4 are highly reliable and don't
>> require redundancy. Attitude control systems, however, are probably on
>> the opposite end of the spectrum and require at least 3 legs, with the
>> standard in the aerospace industry actually using 4 computers (Boeing
>> 777, Joint Strike Fighter, etc.). Without us (NASA) knowing where
>> redundancy is and isn't required, we place the burden on the contractor
>> to decide for us.
>>
>>   Since redundancy is an integral part of the overall design, any
>> changes to redundancy after a basic design is presented could
>> potentially significantly impact the basic design, usually at a monetary
>> cost, which will then influence the overall decision-making process.
>>
>>   I believe it would be worthwhile for us (NASA) to take a hard
>> look historically at which systems, components, etc., have been very
>> reliable and therefore don't require redundancy, and which systems
>> require redundancy in order to achieve the necessary reliability BEFORE
>> a basic design is given to us. This will allow us to better direct the
>> contractor in their efforts to build a highly reliable system, and it
>> will allow us to better assess whether or not the contractor has, in
>> fact, built a highly reliable system.
>>
>>   I'm not sure if you're the right guy to go to or not. My
>> feeling was that ideally this was something that should be done by an
>> "independent" group of folks rather than allowing the program to do
>> it. A look at the standards in the industry, which components and
>> industries historically have been highly reliable and which haven't,
>> would be very helpful for both us and the contractor for the next
>> vehicle. If you could let me know what you think, I would appreciate
>> it. Like I said, I'm not sure if you're the right guy, or if I should
>> run this by someone else, or if the project would be to big to undertake
>> and it's just easier to look at the final design, but my gut feel is
>> that this may be worth doing.
>>
>>   Hope all is well in VA. We loved living there this time of year
>> when all the leaves where changing the cold fronts were coming through.
>>
>> Bloomer
>>
>

Form Approval and Document Revision History

Approved: _____
NESC Director   Date

Version	Description of Revision	Office of Primary Responsibility	Effective Date
1.0	Initial Release	Principal Engineers Office	29 Jan 04

NESC Request Form
NESC-PR-003-FM-01, v1.0

NESC Request Number: 05-173-E

Appendix B. Team List

Name	Position/Role	Center/Affiliation
Mr. Jay Leggett	Team Lead	JSC
Mrs. Julie Kramer White	Team Lead (Former)	JSC
Mr. James Miller	Deputy Team Lead	LaRC
Dr. Bernard Adelstein	Human Factors	ARC
Mr. Michael Aguilar	Software	GSFC
Mr. Michael Bay	SE/Avionics Lead	Bay Engineering
Dr. Michael Cleary	SE	Draper Laboratory
Mr. Mitchell Davis	Electronics Lead	GSFC
Mr. Cornelius Dennehy	GN&C Lead	GSFC
Mr. Gerald Gilmore	SE	Draper Laboratory
Mr. George Hopson	Propulsion Lead	MSFC
Mr. John McManamen	Mechanisms Lead	JSC
Dr. Bob Piascik	Materials Lead	LaRC
Dr. Dorothy Poppe	SE	Draper Laboratory
Dr. Ivatury Raju	Structures Lead	LaRC
Mr. Hank Rotter	Life Support and Active Thermal Lead	JSC
Mr. Allan Cohen	SE/History	The Aerospace Corp.
Mr. Blake Putney	SE/Reliability	Valador
Mr. Robert Ryan	SE	MSFC
Mr. Luke Schutzenhofer	SE	MSFC
Mr. James Blair	SE	MSFC
Support		
Ms. Kim Cannon	Program Analyst	LaRC
Ms. Christina Cooper	Project Coordinator	Swales Aerospace, LaRC
Mrs. Stacey Walker/ Ms. Christina Cooper	Technical Writers	Swales Aerospace, LaRC
Independent Reviewers and Contributors		
Mr. TK Mattingly	Independent Reviewer and Contributor	Systems Planning and Analysis

NESC Request Number: 05-173-E

www.ingramcontent.com/pod-product-compliance
Lightning Source LLC
Chambersburg PA
CBHW081726170526
45167CB00009B/3720